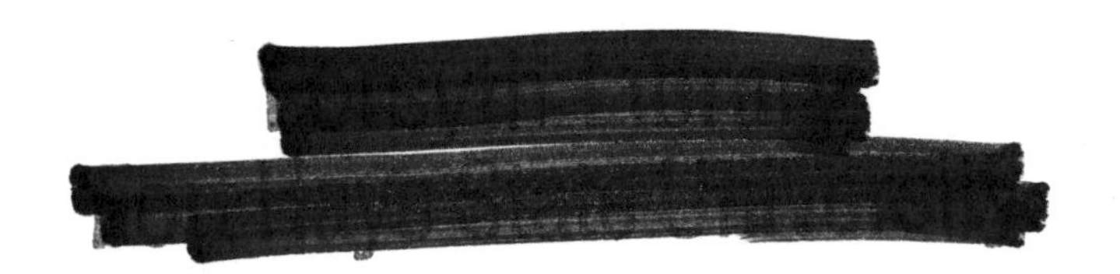

Lax-Phillips Scattering and Conservative Linear Systems: A Cuntz-Algebra Multidimensional Setting

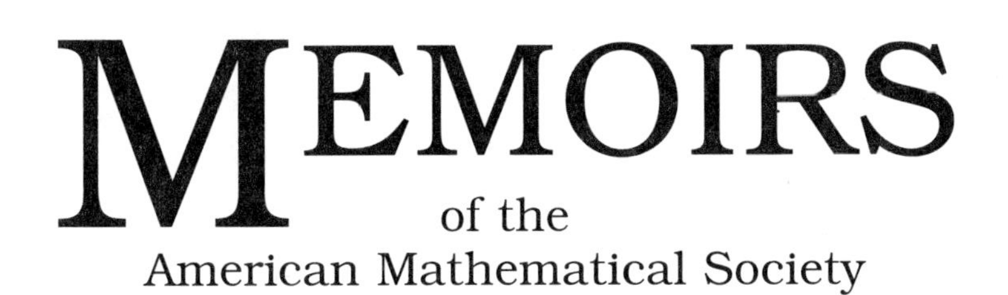

Number 837

Lax-Phillips Scattering and Conservative Linear Systems: A Cuntz-Algebra Multidimensional Setting

Joseph A. Ball
Victor Vinnikov

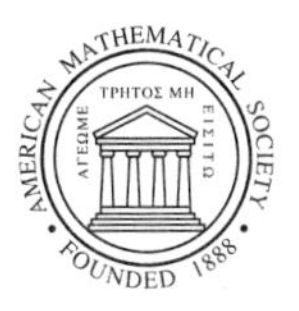

November 2005 • Volume 178 • Number 837 (first of 5 numbers) • ISSN 0065-9266

American Mathematical Society
Providence, Rhode Island

2000 *Mathematics Subject Classification.*
Primary 47A48; Secondary 13F25, 47A40, 47L30, 47L55, 93C05.

Library of Congress Cataloging-in-Publication Data

Ball, Joseph A., 1947–
Lax-Phillips scattering and conservative linear systems: A Cuntz-algebra multidimensional setting / Joseph A. Ball, Victor Vinnikov.
p. cm. — (Memoirs of the American Mathematical Society, ISSN 0065-9266 ; no. 837)
"Volume 178, number 837 (first of 5 numbers)."
Includes bibliographical references.
ISBN 0-8218-3768-0 (alk. paper)
1. Scattering (Mathematics). 2. Linear systems. 3. Operator algebras. 4. Hilbert space.
I. Vinnikov, V. (Victor). II. Title. III. Series.

QA3.A57 no. 837
[QA402]
510 s—dc22
[515′.724] 2005051349

Memoirs of the American Mathematical Society

This journal is devoted entirely to research in pure and applied mathematics.

Subscription information. The 2005 subscription begins with volume 173 and consists of six mailings, each containing one or more numbers. Subscription prices for 2005 are $606 list, $485 institutional member. A late charge of 10% of the subscription price will be imposed on orders received from nonmembers after January 1 of the subscription year. Subscribers outside the United States and India must pay a postage surcharge of $31; subscribers in India must pay a postage surcharge of $43. Expedited delivery to destinations in North America $35; elsewhere $130. Each number may be ordered separately; *please specify number* when ordering an individual number. For prices and titles of recently released numbers, see the New Publications sections of the *Notices of the American Mathematical Society.*

Back number information. For back issues see the *AMS Catalog of Publications.*

Subscriptions and orders should be addressed to the American Mathematical Society, P. O. Box 845904, Boston, MA 02284-5904, USA. *All orders must be accompanied by payment.* Other correspondence should be addressed to 201 Charles Street, Providence, RI 02904-2294, USA.

Memoirs of the American Mathematical Society is published bimonthly (each volume consisting usually of more than one number) by the American Mathematical Society at 201 Charles Street, Providence, RI 02904-2294, USA. Periodicals postage paid at Providence, RI. Postmaster: Send address changes to Memoirs, American Mathematical Society, 201 Charles Street, Providence, RI 02904-2294, USA.

This publication is indexed in *Science Citation Index®*, *SciSearch®*, *Research Alert®*, *CompuMath Citation Index®*, *Current Contents®/Physical, Chemical & Earth Sciences.*
Printed in the United States of America.

∞ The paper used in this book is acid-free and falls within the guidelines
established to ensure permanence and durability.
Visit the AMS home page at http://www.ams.org/

10 9 8 7 6 5 4 3 2 1 10 09 08 07 06 05

Contents

Abstract

We present a multivariable setting for Lax-Phillips scattering and for conservative, discrete-time, linear systems. The evolution operator for the Lax-Phillips scattering system is an isometric representation of the Cuntz algebra, while the nonnegative time axis for the conservative, linear system is the free semigroup on d letters. The correspondence between scattering and system theory and the roles of the scattering function for the scattering system and the transfer function for the linear system are highlighted. Another issue addressed is the extension of a given representation of the Cuntz-Toeplitz algebra (i.e., a row isometry) to a representation of the Cuntz algebra (i.e., a row unitary); the solution to this problem relies on an extension of the Szegö factorization theorem for positive Toeplitz operators to the Cuntz-Toeplitz algebra setting. As an application, we obtain a complete set of unitary invariants (the characteristic function together with a choice of "Haplitz" extension of the characteristic function defect) for a row-contraction on a Hilbert space.

Received by the editor November 13, 2002 and in revised form September 28, 2004.

2000 *Mathematics Subject Classification.* Primary: 47A48; Secondary: 13F25, 47A40, 47L30, 47L55, 93C05.

Key words and phrases. formal power series, noncommuting indeterminants, unitary colligation, incoming and outgoing space, scattering function, row isometry, row contraction, functional model.

The first author is supported by NSF grant DMS-998763; both authors are supported by a grant from the US-Israel Binational Science Foundation.

CHAPTER 1

Introduction

It is well known that unitary linear system theory, scattering theory, and operator model theory are all closely related; for good explanations of various facets of these theories and of their connections with each other, we refer to [**Bro71, Li73, NaF70, BraR66, BaC91, ADRS97, LaP67, AA95, NiV89, NiV98, BaT00**] and the survey article [**Ba00**]. Indeed, there is essentially a one-to-one correspondence between each pair of these objects (once some natural minimality and normalization requirements are imposed). Each theory produces a contractive operator function $W(z)$ on the unit disk (called the transfer function of the linear system, the scattering function of the scattering system, and the characteristic function of the completely nonunitary (c.n.u.) contraction operator, respectively) from which one can recover (up to unitary equivalence) the original object (unitary system, scattering system, c.n.u. contraction operator, respectively). One can produce the one-to-one correspondence between each pair of objects by way of this operator function or via a more direct mapping, the details of which depend on the pair under consideration.

A Lax-Phillips scattering system (as generalized by Adamjan and Arov) amounts to a Hilbert space $\mathcal{K}$ together with a unitary operator $\mathcal{U}$ on $\mathcal{K}$ and two subspaces $\mathcal{G}$ (the *outgoing space*) and $\mathcal{G}_*$ (the *incoming space*), invariant for $\mathcal{U}$ and $\mathcal{U}^*$ respectively, such that $\mathcal{U}|_{\mathcal{G}}$ is a unilateral shift operator with wandering subspace $\mathcal{E} = \mathcal{G} \ominus \mathcal{U}\mathcal{G}$ and $\mathcal{U}^*|_{\mathcal{G}_*}$ is a unilateral shift operator with wandering subspace $\mathcal{U}^*\mathcal{E}_*$ (where $\mathcal{E}_* = \mathcal{U}\mathcal{G}_* \ominus \mathcal{G}_*$). One can then use a natural Fourier representation to represent $\widetilde{\mathcal{G}} =$ closed $\text{span}_{n=1,2,\dots} \mathcal{U}^{*n}\mathcal{G}$ as $L^2(\mathbb{T}, \mathcal{E})$ (the L^2-space of $\mathcal{E}$-valued functions on the unit circle $\mathbb{T}$) with $\mathcal{U}|_{\widetilde{\mathcal{G}}}$ represented as multiplication by the coordinate function z, and with the subspace $\mathcal{G} \subset \widetilde{\mathcal{G}}$ represented as the Hardy subspace $H^2(\mathbb{T}, \mathcal{E})$. Similarly, one can represent $\widetilde{\mathcal{G}}_* =$ closed $\text{span}_{n=1,2,\dots} \mathcal{U}^n\mathcal{G}_*$ as $L^2(\mathbb{T}, \mathcal{E}_*)$ with $\mathcal{U}|_{\widetilde{\mathcal{G}}_*}$ represented as multiplication by z on $L^2(\mathbb{T}, \mathcal{E}_*)$ and with the subspace $\mathcal{G}_* \subset \widetilde{\mathcal{G}}_*$ represented as $H^2(\mathbb{T}, \mathcal{E}_*)^\perp$. The restricted orthogonal projection $P_{\widetilde{\mathcal{G}}_*}|_{\widetilde{\mathcal{G}}}$ then corresponds to a multiplication operator $f(z) \mapsto S(z)f(z)$ where $S(z) \in \mathcal{S}(\mathcal{E}, \mathcal{E}_*)$ (the *Schur class* of holomorphic functions on the unit disk with values equal to contractive operators between $\mathcal{E}$ and $\mathcal{E}_*$) is the *scattering function* for the system. Conversely, given a Schur-class function $S \in \mathcal{S}(\mathcal{E}, \mathcal{E}_*)$, one can construct a model scattering system $\mathfrak{S}_S$ which has S has its scattering function, and any minimal scattering system with scattering function equal to S is unitarily equivalent to its model scattering system $\mathfrak{S}_S$. In this way one sees that the scattering function S is a complete unitary invariant for minimal scattering systems.

A *unitary colligation* amounts to a block unitary operator U of the form

$$U = \begin{bmatrix} A & B \\ C & D \end{bmatrix} : \begin{bmatrix} \mathcal{H} \\ \mathcal{E} \end{bmatrix} \to \begin{bmatrix} \mathcal{H} \\ \mathcal{E}_* \end{bmatrix}$$

to which corresponds a discrete-time, unitary, linear system

$$\Sigma\colon \left\{ \begin{array}{rcl} x(n+1) & = & Ax(n)+Bu(n) \\ y(n) & = & Cx(n)+Du(n). \end{array} \right. \tag{1.1}$$

Here the "time-variable" n is an element of the nonnegative integers $\mathbb{Z}_+$ (or sometimes the set of all integers $\mathbb{Z}$), the vector $x(n)$ takes values in the state space $\mathcal{H}$, the vector $u(n)$ takes values in the input space $\mathcal{E}$, and the vector $y(n)$ takes values in the output space $\mathcal{E}_*$. Application of the Fourier transform $x(z) \mapsto \widehat{x}(z) = \sum_{n=0}^{\infty} x(n)z^n$ to the system equations (1.1), under the assumption of a zero initial condition $x(0) = 0$ on the state vector, leads to the input-output relation

$$\widehat{y}(z) = T_\Sigma(z)\widehat{u}(z)$$

where $T_\Sigma(z)$ is the *transfer function* of the system

$$T_\Sigma(z) = D + zC(I - zA)^{-1}B.$$

When one starts from the point of view of operator-model theory, one is given a contraction operator $\mathbf{T}$ on a Hilbert space $\mathcal{H}$. Associated with $\mathbf{T}$ is a Schur-class function $T_{\mathbf{T}}(z) \in \mathcal{S}(\mathcal{D}, \mathcal{D}_*)$, the *characteristic operator function* for $\mathbf{T}$, where $\mathcal{D}$ and $\mathcal{D}_*$ are the defect spaces associated with $\mathbf{T}$. Given any Schur-class function T, one can then build a model operator $\mathbf{T}_T$ on a functional Hilbert space $\mathcal{H}_{\mathbf{T}}$ so that the characteristic function of $\mathbf{T}_T$ is equal to T, and any completely nonunitary contraction operator $\mathbf{T}$ with characteristic function equal to T is unitarily equivalent to its model operator $\mathbf{T}_T$; in this way we see that the characteristic function $T_{\mathbf{T}}$ is a complete unitary invariant for completely nonunitary contraction operators $\mathbf{T}$ on a Hilbert space $\mathcal{H}$.

Note that the same Schur-class function $S \in \mathcal{S}(\mathcal{E}, \mathcal{E}_*)$ can serve as the scattering function for a Lax-Phillips system, as the transfer function for a conservative, discrete-time, input-state-output linear system $\Sigma(U)$, and as the characteristic operator function for a contraction operator $\mathbf{T}$. Thus there is a natural association and equivalence between Lax-Phillips scattering systems $\mathfrak{S}$, conservative, discrete-time, input-state-output linear systems $\Sigma(U)$, and Hilbert-space contraction operators $\mathbf{T}$.

In the generalization which we discuss here, the unitary operator $\mathcal{U}$ in a Lax-Phillips scattering system $\mathfrak{S}$ is replaced by a d-tuple of isometries $\mathcal{U} = (\mathcal{U}_1, \dots, \mathcal{U}_d)$ on a Hilbert space $\mathcal{K}$ such that the block row $\begin{bmatrix} \mathcal{U}_1 & \dots & \mathcal{U}_d \end{bmatrix} \colon \oplus_{k=1}^d \mathcal{K} \to \mathcal{K}$ is unitary, i.e., $(\mathcal{U}_1, \dots, \mathcal{U}_d)$ provides an isometric representation of the Cuntz algebra $\mathcal{O}_d$ on $\mathcal{K}$ (see e.g. the book of Davidson [**Da96**] for basic definitions). To set up a scattering framework for this situation we hypothesize the existence of subspaces $\mathcal{G}$ and $\mathcal{G}_*$ such that $\mathcal{G}$ and $\mathcal{G}_*$ are invariant for $\mathcal{U}_1, \dots, \mathcal{U}_d$ and $\mathcal{U}_1^*, \dots, \mathcal{U}_d^*$ respectively, and $\mathcal{U}|_{\mathcal{G}}$ is a "row shift" and $\mathcal{U}^*|_{\mathcal{G}_*}$ is the "backwards row shift" (precise definitions are given in Chapter 3). The Fourier representation theory is now more complicated, but it turns out that one can represent the smallest closed subspace $\widetilde{\mathcal{G}}$ containing $\mathcal{G}$ and reducing for each of the operators $\mathcal{U}_1, \dots, \mathcal{U}_d$ as a certain functional model space $\mathcal{L}_W$ for Cuntz-algebra representations, with $\mathcal{G}$ corresponding to a certain analogue $\mathcal{H}_W$ of a Hardy subspace. Similarly, the smallest closed subspace $\widetilde{\mathcal{G}}_*$ containing $\mathcal{G}_*$ and reducing for each $\mathcal{U}_1, \dots, \mathcal{U}_d$ can be represented as a Cuntz-algebra functional model space $\mathcal{L}_{W_*}$, with $\mathcal{G}_*$ corresponding to $\mathcal{H}_{W_*}^{\perp}$. Then the restricted projection operator $P_{\widetilde{\mathcal{G}}_*}|_{\widetilde{\mathcal{G}}}$ corresponds to a certain multiplication operator $f \mapsto S \cdot f$ where S now is a certain formal power series in two d-tuples of noncommuting variables belonging to a noncommutative version $\mathcal{S}(W, W_*)$ of the Schur-class. As before, this $S \in \mathcal{S}(W, W_*)$ is a complete unitary invariant for such Cuntz scattering systems.

In our companion paper [**BaV04**] we obtained a classification of isometric representations of the Cuntz algebra where one in addition specifies a choice of cyclic subspace $\mathcal{E}$. The classification is in terms of a block matrix $[W_{v,w;\alpha,\beta}]$ with rows and columns indexed by the $\mathcal{F}_d \times \mathcal{F}_d$ (where $\mathcal{F}_d$ is the free semigroup on d letters) possessing a combination of Toeplitz and Hankel properties; for this reason we shall call such matrices "Haplitz matrices". In another report [**BaV03**], we identify this model as an example of a *noncommutative formal reproducing kernel Hilbert space*, a Hilbert space of formal power series in finitely many noncommuting variables having many properties analogous to classical reproducing kernel Hilbert spaces. This is one of the main tools for the discussion here, and will be reviewed (along with some additional needed results) in the preliminary Chapter 2.

The Cuntz analogue of a unitary colligation is as follows. A *d-variable unitary colligation* is a unitary operator of the form

$$U = \begin{bmatrix} A & B \\ C & D \end{bmatrix} = \begin{bmatrix} A_1 & B_1 \\ \vdots & \vdots \\ A_d & B_d \\ C & D \end{bmatrix} : \begin{bmatrix} \mathcal{H} \\ \mathcal{E} \end{bmatrix} \to \begin{bmatrix} \oplus_1^d \mathcal{H} \\ \mathcal{E}_* \end{bmatrix}. \tag{1.2}$$

The positive time axis $\mathbb{Z}^+$ is replaced by a free semigroup $\mathcal{F}_d$ with generators equal to the d letters $g_1, \dots, g_d$, and the associated $\mathcal{F}_d$-time system is given by

$$\Sigma\colon \left\{ \begin{array}{rcl} x(g_1 w) & = & A_1 x(w) + B_1 u(w) \\ & \vdots & \\ x(g_d w) & = & A_d x(w) + B_d u(w) \\ y(w) & = & Cx(w) + Du(w) \end{array} \right. \tag{1.3}$$

where the variable $w = g_{i_n} \cdots g_{i_1}$ is a word in the symbols $g_1, \dots, g_d$, i.e., a generic element of the free semigroup $\mathcal{F}_d$. The Fourier transform in this context we take to be $x(w) \mapsto \widehat{x}(z) = \sum_{w \in \mathcal{F}_d} x(w) z^w$ where $z^w = z_{i_n} \cdots z_{i_1}$ if $w = g_{i_n} \cdots g_{i_1}$ where $z = (z_1, \dots, z_d)$ is a d-tuple of formal, noncommuting indeterminants. Upon application of this generalized Fourier transform to the system equations (1.3) and under the assumption that the state of the system is initialized at 0 (so $x(\emptyset) = 0$ where $\emptyset$ is the empty word, the unit element of $\mathcal{F}_d$), we arrive at the input-output relation

$$\widehat{y}(z) = W_\Sigma(z) \widehat{u}(z)$$

where the *transfer function* in this case is given by

$$T_\Sigma(z) = D + C(I - Z_r(z)A)^{-1} Z_r(z) B$$

where we have set $Z_r(z)$ equal to the block-row matrix function

$$Z_r(z) = \begin{bmatrix} z_1 I_{\mathcal{H}} & \dots & z_d I_{\mathcal{H}} \end{bmatrix}.$$

To make connections with the scattering framework, it is also necessary to run the system in backwards time; this is more complicated and is described in Chapter 3. We mention that such formal power series have come up in various contexts in the system theory literature (see e.g. [**Fl74, Be01, BeD99**]).

There is also a dilation and operator model theory which fits into this setting. Rather than considering a single contraction operator $\mathbf{T}$ on a HIlbert space

$\mathcal{H}$, we consider a d-tuple $\mathbf{T} = (\mathbf{T}_1, \dots, \mathbf{T}_d)$ of operators on a Hilbert space for which the block row operator $\begin{bmatrix} \mathbf{T}_1 & \dots & \mathbf{T}_d \end{bmatrix} : \oplus_{k=1}^d \mathcal{H} \to \mathcal{H}$ is a contraction (a *row contraction*); one obtains a Sz.-Nagy-Foiaş-type operator model by analyzing the geometry of the space of its minimal row-isometric dilation (see [**Fr82, Bu84, Po89a, Po89b**]); in particular Popescu [**Po89b**] has shown that *the characteristic function for a row contraction is a complete unitary invariant in case the row contraction is "completely noncoisometric"*. In addition, we now have a Wold decomposition for a row isometry [**Fr84, Po95**] together with a Beurling-Lax Theorem [**Po89b, DaP99**] and Sz.-Nagy-Foiaş Commutant Lifting Theorem for this setting [**Po89c, Po95**], a von Neumann inequality [**Po99**], a noncommutative analogue of the C^*-algebra of analytic Toeplitz operators [**DaP98a**], and a noncommutative analogue of Nevanlinna-Pick interpolation [**DaP98b, Po98, AP00, CSK02**]. When one symmetrizes the underlying Fock space, one arrives at a reproducing kernel Hilbert space $\mathcal{H}(k_d)$ of analytic functions over the unit ball in $\mathbb{C}^d$ and its associated space of bounded multipliers for which many parallel results hold (see [**Dr78, Po91, Arv98, AM00, BaTV01, McT00**]). Moreover, we now have a fairly complete understanding of a certain partial invariant derived from the characteristic function, namely the "curvature invariant" (see [**Kr01, Po01a**] for the general setting and [**Arv00, GRS02**] for the commutative setting where the idea actually originated).

The main focus of the paper is to lay out the connections between scattering, conservative linear systems and operator model theory for contractions in the Cuntz-algebra setting. The following points summarize our main results:

(1) *A Cuntz scattering system $\mathfrak{S}$, if minimal, is determined up to unitary equivalence by its scattering operator $S_{\mathfrak{S}}$, and is unitarily equivalent to any of three model Cuntz scattering systems built from its scattering function.*

(2) *A Cuntz scattering system $\mathfrak{S}$ determines a unique Cuntz unitary colligation $U = U(\mathfrak{S})$; the transfer function $T_{\Sigma(U)}$ for the associated Cuntz conservative linear system $\Sigma(U)$ is a certain restriction of the scattering function $S_{\mathfrak{S}}$ of the Cuntz scattering system.*

(3) *Conversely, a Cuntz unitary colligation U, together with a certain additional invariant W_* (a "shift-like Cuntz weight" on output signals of the system $\Sigma(U)$) uniquely determines a Cuntz scattering system $\mathfrak{S}(U, W_*)$. If the Cuntz unitary colligation U has the form $U = U(\mathfrak{S})$ for a Cuntz scattering system $\mathfrak{S}$, then there is a particular choice of weight $W_* = W_*(\mathfrak{S})$ determined by $\mathfrak{S}$ so that we recover $\mathfrak{S}$ as $\mathfrak{S} = \mathfrak{S}(U, W_*)$. Moreover, the outgoing shift Cuntz weight W and the scattering operator S for $\mathfrak{S}(U, W_*)$ associated with $\mathfrak{S}(U, W_*)$ can be given explicitly in terms of U and W_*.*

(4) *A Cuntz unitary colligation uniquely determines its characteristic function $T(z) = \sum_{v \in \mathcal{F}_d} T_v z^v$ (a formal power series in the noncommutative, d-variable Schur-class $\mathcal{S}_{nc,d}(\mathcal{E}, \mathcal{E}_*)$—see (2.4.1) for the precise definition) as well as a particular choice of "Haplitz" extension L of the Cuntz-Toeplitz defect operator $I - (M_T)^* M_T$. Moreover, the colligation can be recovered from (T, L) up to unitary equivalence from a Sz.-Nagy-Foiaş model constructed from (T, L); hence (T, L) forms a complete set of unitary invariants for a closely-connected Cuntz-unitary colligation.*

(5) *By the Halmos dilation process, a completely nonunitary row contraction* $\mathbf{T} = (\mathbf{T}_1, \dots, \mathbf{T}_d)$ *uniquely determines a "strict", closely-connected Cuntz unitary colligation* U (1.2) *(with* $(\mathbf{T}_1, \dots, \mathbf{T}_d) = (A_1^*, \dots, A_d^*)$*). Hence a complete set of unitary invariants is the pair* (T, L) *where* T *is the characteristic function for* $\mathbf{T}$ *(i.e., the characteristic function for the colligation* U*) and* L *is a choice of Haplitz extension of the Cuntz-Toeplitz defect operator* $I - (M_T)^* M_T$*. In fact, the row contraction* $\mathbf{T}$ *is unitarily equivalent to its Sz.-Nagy-Foiaş model row contraction* $\mathbf{T}_{(T,L)}$ *constructed from its characteristic pair* (T, L). This removes the "completely noncoisometric" restriction in the result of Popescu cited above.

The paper is organized as follows. After the present Introduction, in Chapter 2 we recall the formalism from [**BaV04**] giving an analogue of L^2-spaces which serves as a model for row unitary operators. We then study in some detail the analogues of the Hardy space $\mathcal{H}_W$ and $\mathcal{H}_W^\perp$ for this setting, as well as the analogues of L^∞ and H^∞, namely, the space of intertwining maps L_T between two row-unitary model spaces $\mathcal{L}_W$ and $\mathcal{L}_{W_*}$, and the subclass of such maps ("analytic intertwining operators") which preserve the associated Hardy spaces ($M_T \colon \mathcal{H}_W \to \mathcal{H}_{W_*}$). Of particular interest is to understand how to extend an intertwining map $M_T \colon \mathcal{H}_W \to \mathcal{H}_{W_*}$ defined only on a Hardy space to a full intertwining map $L_T \colon \mathcal{L}_W \to \mathcal{L}_{W_*}$. The contractive, analytic intertwining operators then form an interesting noncommutative analogue of the "Schur class" which has been receiving much attention of late from a number of points of view (see e.g. [**Ba00**]). Preliminary to this analysis is understanding how to extend a (noncommutative) Hardy space $\mathcal{H}_{W^+}$ to a full noncommutative Lebesgue space $\mathcal{L}_W$, or equivalently, how to extend a given noncommutative Toeplitz operator W^+ to a full Haplitz operator W. An equivalent operator-theoretic formulation is how to extend a given row isometry to a row unitary; unlike as in the classical case, this extension generally is not unique. Our analysis of these issues is via a noncommutative analogue of the Szegö factorization theory for a nonnegative Toeplitz operator; here there is overlap with the seminal work of Popescu (see [**Po01b**]), who also gives noncommutative analogues of the connections with prediction theory and entropy optimization, and with the work of Adams, Froelich, McGuire and Paulsen (see [**AFMP94**]), who discuss such factorization results in a commutative but non-Toeplitz setting.

In Chapter 3, we axiomatize the notion of a "Cuntz Lax-Phillips scattering system", define the scattering function for such an object and the associated functional models (of Pavlov, de Branges-Rovnyak and Sz.-Nagy-Foiaş type) for any minimal Cuntz scattering system. Chapter 4 introduces the multidimensional system Σ and unitary colligation corresponding to a Cuntz Lax-Phillips scattering system, along with the associated transfer function characterizing input-output properties and the mappings back and forth between Cuntz scattering systems and Cuntz unitary colligations. The final Chapter 5 presents the results on unitary classification of Cuntz unitary colligations, or equivalently, of completely nonunitary row contractions. Here we also explain how the curvature invariant $\operatorname{curv}(\mathbf{T})$ for a row contraction $\mathbf{T} = (\mathbf{T}_1, \dots, \mathbf{T}_d)$ with finite-rank defect operator $D_{\mathbf{T}}$ introduced in [**Kr01**] and [**Po01a**] can be computed directly in terms of the characteristic function $T_{\mathbf{T}}(z)$ of $\mathbf{T}$.

CHAPTER 2

Functional Models for Row-Isometric/Row-Unitary Operator Tuples

2.1. The classical case

If $\mathcal{K}$ is a Hilbert space, $\mathcal{E}$ is a subspace of $\mathcal{K}$ and $\mathcal{U}$ is a unitary operator on $\mathcal{K}$, we may associate a block-Toeplitz matrix $[W_{i,j}]_{i,j=-\infty}^{\infty}$ by

$$W_{i,j} = P_{\mathcal{E}}\mathcal{U}^{*i}\mathcal{U}^{j}|_{\mathcal{E}} = P_{\mathcal{E}}\mathcal{U}^{j-i}|_{\mathcal{E}}.$$

Note that each matrix entry is an operator on $\mathcal{E}$. It is convenient to view W as an operator from the space $\mathcal{P}(\mathbb{Z}, \mathcal{E})$ of trigonometric polynomials $p(z) = \sum_{n=-N}^{N} p_n z^n$ with coefficients $p_n \in \mathcal{E}$ into the space $L(\mathbb{Z}, \mathcal{E})$ of formal Laurent series $f(z) = \sum_{n=-\infty}^{\infty} f_n z^n$ with coefficients $f_n \in \mathcal{E}$ via the formula

$$f(z) = (Wp)(z) \text{ if } f_i = \sum_{j=-\infty}^{\infty} W_{i,j}p_j = \sum_{j=-N}^{N} W_{i,j}p_j.$$

From the factored form of $W_{i,j}$, it is easily seen that $W_{i,j}$ is *positive semidefinite* in the sense that

$$\langle Wp, p\rangle_{L^2} \geq 0 \text{ for all } p \in \mathcal{P}(\mathbb{Z}, \mathcal{E}.$$

Here we are using that the usual L^2-inner product

$$\left\langle \sum_{n=-\infty}^{\infty} f_n z^n, \sum_{n=-\infty}^{\infty} p_n z^n \right\rangle_{L^2} = \sum_{n=-\infty}^{\infty} \langle f_n, p_n\rangle_{\mathcal{E}}$$

gives a well-defined pairing between $L(\mathbb{Z}, \mathcal{E})$ and $\mathcal{P}(\mathbb{Z}, \mathcal{E})$, since the infinite sum for such a pairing collapses to a finite sum. The image $W \cdot \mathcal{P}(\mathbb{Z}, \mathcal{E})$ of this operator W can be completed to a Hilbert space denoted by $\mathcal{L}_W$ with inner product defined by

$$\langle Wp, Wq\rangle_{\mathcal{L}_W} = \langle Wp, q\rangle_{L^2}. \tag{2.1.1}$$

As evaluation of Fourier coefficients $f(z) \mapsto f_n$ is bounded in this norm, elements of the completion can still be identified as formal series $f(z) = \sum_{n=-\infty}^{\infty} f_n z^n$. Moreover, the map Φ given by

$$\Phi \colon k \mapsto \sum_{n=-\infty}^{\infty} (P_{\mathcal{E}}\mathcal{U}^{*n}k)z^n, \tag{2.1.2}$$

sometimes called the *Fourier representation operator* with respect to $\mathcal{U}$ and $\mathcal{E}$, defines a coisometry from $\mathcal{K}$ onto $\mathcal{L}_W$ such that

$$\Phi\mathcal{U} = \mathcal{U}_W\Phi$$

where $\mathcal{U}_W \colon \mathcal{L}_W \to \mathcal{L}_W$ is the operator of multiplication by the indeterminant z on $\mathcal{L}_W$:

$$\mathcal{U}_W \colon f(z) \mapsto zf(z),$$

and we recover W as $W = \Phi\Phi^{[*]}$ (where $\Phi^{[*]}$ is the formal adjoint of Φ with respect to the $\mathcal{K}$-inner product on its domain and the formal L^2-inner product on its range). In particular, $\mathcal{U}_W$ is unitary on $\mathcal{L}_W$, and, if $\mathcal{E}$ is $*$-cyclic for $\mathcal{U}$, then Φ implements a unitary equivalence between $\mathcal{U}$ on $\mathcal{K}$ and the functional model $\mathcal{U}_W$ on $\mathcal{L}_W$. Conversely, if W is *any* positive-semidefinite block-Toeplitz matrix, we may define a space $\mathcal{L}_W$ as above, and an operator $\mathcal{U}_W \colon f(z) \mapsto zf(z)$ on $\mathcal{L}_W$. The Toeplitz structure of W guarantees that $\mathcal{L}_W$ is unitary. It is often more convenient to work with the *symbol* of W, the formal Laurent series $\widehat{W}(z) = \sum_{n=-\infty}^{\infty} \widehat{W}_n z^n$ with operator coefficients determined by $W(z) \cdot e = (We)(z)$ for $e \in \mathcal{E}$ (considered as a constant function in $\mathcal{P}(\mathbb{Z}, \mathcal{E})$). Equivalently, we have $W_{i,j} = \widehat{W}_{i-j}$. If one interprets the formal indeterminate z as a variable inside the unit disk $\mathbb{D}$, the formula $\widehat{W}(z) = \sum_{n=-\infty}^{\infty} \widehat{W}_n z^n$ defines a harmonic function on the unit disk which therefore has a Poisson representation

$$\widehat{W}(z) = \frac{1}{2\pi} \int_{\mathbb{T}} \frac{1 - |z|^2}{|1 - z\overline{\tau}|^2}\, d\sigma(\tau)$$

for a positive operator-valued measure τ on $\mathbb{T}$. Some authors prefer to work with the analytic function with positive-real part on the unit disk having the Herglotz representation

$$\varphi(z) = \sum_{n=0}^{\infty} \widehat{W}_n z^n = \frac{1}{2\pi} \int_{\mathbb{T}} \frac{\tau + z}{\tau - z}\, d\sigma(\tau).$$

Closely related functional models for unitary operators have appeared in the literature (see [**BoDK00**] and [**BraR66**]). Indeed, the space $\mathcal{L}_W$ can be identified as the space of all formal power series $f(z) = \sum_{n=-\infty}^{\infty} f_n z^n$ such that $f_n = \frac{1}{2\pi} \int_{\mathbb{T}} \tau^{-n}\, dv(\tau)$ where the vector measure v (called a *chart* in the terminology of [**BoDK00**]) is in the Hellinger space $\mathcal{L}^{\sigma}$ delineated in [**BoDK00**]. Alternatively, $f(z) = \sum_{n=-\infty}^{\infty} f_n z^n \in \mathcal{L}_W$ if and only if the pair of analytic functions $(\widetilde{f}(z), \widetilde{g}(z))$ defined by $\widetilde{f}(z) = \sum_{n=0}^{\infty} f_n z^n$ and $\widetilde{g}(z) = \sum_{n=0}^{\infty} f_{-n-1} z^n$ for z in the unit disk $\mathbb{D}$ is in the space $\mathcal{E}(\varphi)$ introduced in [**BraR66**].

2.2. Functional models of noncommutative formal power series

The paper [**BaV04**] presents an extension of these ideas to the setting where the single unitary operator $\mathcal{U}$ is replaced by a row-unitary operator-tuple $\mathcal{U} = (\mathcal{U}_1, \dots, \mathcal{U}_d)$. Given a d-tuple of operators $\mathcal{U} = (\mathcal{U}_1, \dots, \mathcal{U}_d)$ on a Hilbert space $\mathcal{K}$, in general we say that $\mathcal{U}$ is a *row isometry* (respectively, *row unitary*) if the operator matrix

$$\begin{bmatrix} \mathcal{U}_1 & \dots & \mathcal{U}_d \end{bmatrix} : \begin{bmatrix} \mathcal{K} \\ \vdots \\ \mathcal{K} \end{bmatrix} \to \mathcal{K}$$

is isometric (respectively, unitary) as an operator from $\oplus_{j=1}^{d} \mathcal{K}$ to $\mathcal{K}$. More concretely, $\mathcal{U}$ is row-isometric if each of $\mathcal{U}_j$ is an isometry, and the image spaces $\operatorname{im} \mathcal{U}_j$ are pairwise orthogonal for $j = 1, \dots, d$. The d-tuple $\mathcal{U}$ is row-unitary if $\mathcal{U}$ is a row-isometry such that the span of the images $\operatorname{im} \mathcal{U}_j$ over $j = 1, \dots, d$ is the whole space $\mathcal{K}$. From the point of view of operator algebras, a row unitary d-tuple $\mathcal{U} = (\mathcal{U}_1, \dots, \mathcal{U}_d)$ amounts to a representation of the Cuntz algebra $\mathcal{O}_d$ (see e.g. [**Da96**] for definitions and further details). For any d-tuple $\mathcal{U} = (\mathcal{U}_1, \dots, \mathcal{U}_d)$ of operators on a Hilbert space $\mathcal{K}$, it is convenient to introduce a functional calculus with respect to the free semigroup $\mathcal{F}_d$ on d letters $g_1, \dots, g_d$. Namely, if $v = g_{i_n} \cdots g_{i_1}$ is a word in $\mathcal{F}_d$ (where $i_n, \dots, i_1 \in \{1, \dots, d\}$), we define $\mathcal{U}$ to the power v, denoted as $\mathcal{U}^v$, by

$$\mathcal{U}^v = \mathcal{U}_{i_n} \dots \mathcal{U}_{i_1}.$$

The d-tuple $\mathcal{U}^*$ is defined as $\mathcal{U}^* = (\mathcal{U}_1^*, \dots, \mathcal{U}_d^*)$. We shall often work with products of the form $\mathcal{U}^w \mathcal{U}^{*v}$. Note that if $\mathcal{U} = (\mathcal{U}_1, \dots, \mathcal{U}_d)$ is row-unitary, then an expression of the form $\mathcal{U}^{*v}\mathcal{U}^w$ collapses according to the following formula which will require some additional explanation:

$$\mathcal{U}^{*v}\mathcal{U}^w = \begin{cases} \mathcal{U}^{*v(w^\top)^{-1}} & \text{if } |v| \geq |w| \\ \mathcal{U}^{(v^\top)^{-1}w} & \text{if } |w| \geq |w|. \end{cases} \tag{2.2.1}$$

Here $|v|$ refers to the *length* of the word v ($|v| = n$ if $v = g_{i_n} \cdots g_{i_1}$), $v^\top$ refers to the *transpose* of v ($v^\top = g_{i_1} \cdots g_{i_n}$ if $v = g_{i_n} \cdots g_{i_1}$) and v^{-1} refers to the *inverse* of v ($v^{-1} = g_{i_1}^{-1} \cdots g_{i_n}^{-1}$ if $v = g_{i_n} \cdots g_{i_1}$). However we do not interpret products of the form $v^{-1}w$ in the free group generated by $(g_1, \dots, g_d)$; rather we define

$$g_\ell^{-1} g_k = \begin{cases} \emptyset & \text{if } \ell = k \\ \text{undefined} & \text{if } \ell \neq k \end{cases}$$

where $\emptyset$ is the empty word of zero length equal to the unit element for $\mathcal{F}_d$. For an operator d-tuple, we interpret $\mathcal{U}^\emptyset = I$ and $\mathcal{U}^{\text{undefined}} = 0$. This completes the interpretation of the formula (2.2.1).

Suppose now that we are given a row-unitary d-tuple $\mathcal{U} = (\mathcal{U}_1, \dots, \mathcal{U}_d)$ on a Hilbert space $\mathcal{K}$ together with some fixed subspace $\mathcal{E}$ of $\mathcal{K}$. We may then associate a matrix $W = [W_{v,w;\alpha,\beta}]$ with rows indexed by elements (v, w) of $\mathcal{F}_d \times \mathcal{F}_d$ and columns indexed by elements (α, β) of $\mathcal{F}_d \times \mathcal{F}_d$ given by

$$W_{v,w;\alpha,\beta} = P_\mathcal{E} \mathcal{U}^w \mathcal{U}^{*v} \mathcal{U}^{\alpha^\top} \mathcal{U}^{*\beta^\top}|_\mathcal{E}. \tag{2.2.2}$$

It is straightforward to verify that this matrix has a *Haplitz structure* (see [**BaV04**]), namely,

$$W_{\emptyset,w;\alpha g_j,\beta} = W_{\emptyset,wg_j;\alpha,\beta}, \tag{2.2.3}$$

$$W_{v,w;\alpha g_j,\beta} = W_{vg_j^{-1},w;\alpha,\beta} \tag{2.2.4}$$

as well as what we call the *Cuntz* property

$$W_{\emptyset,w;\emptyset,\beta} = \sum_{j=1}^{d} W_{\emptyset,wg_j;\emptyset,\beta g_j}. \tag{2.2.5}$$

Note that (2.2.3) is a *Hankel*-like property while (2.2.4) is a *Toeplitz*-like property (and hence the term *Haplitz*). Property (2.2.5) is equivalent to the row unitary property of $\mathcal{U}$ (i.e., $\sum_{j=1}^d \mathcal{U}_j \mathcal{U}_j^* = I_\mathcal{K}$), i.e., to $\mathcal{U}$ inducing a representation of the *Cuntz* algebra. Given that $W = W^{[*]}$ (where $[*]$ denotes the conjugate transpose of W), we also have

$$\begin{aligned} W_{v,w;\emptyset,\beta g_j} &= [W_{\emptyset,\beta g_j;v,w}]^* \\ &= [W_{\emptyset,\beta;vg_j,w}]^* \\ &= W_{vg_j,w;\emptyset,\beta} \end{aligned}$$

and hence

$$W_{vg_j,w;\emptyset,\beta} = W_{v,w;\emptyset,\beta g_j}. \tag{2.2.6}$$

In parallel with our introductory discussion of the case $d = 1$, let us think of W as an operator acting on certain formal function spaces. For this purpose, we let $z = (z_1, \dots, z_d)$ and $\zeta = (\zeta_1, \dots, \zeta_d)$ be two sets of noncommuting indeterminants. If $v = g_{i_n} \cdots g_{i_1}$ is a word in $\mathcal{F}_d$, we define z^v to be $z_{i_n} \dots z_{i_1}$ and similarly for ζ^v. The rules are that z_i's do not commute with each other, ζ_j's do not commute with each other, but we do have $z_i\zeta_j = \zeta_j z_i$ for $i, j = 1, \dots, d$. We denote by $L(\mathcal{F}_d \times \mathcal{F}_d, \mathcal{E})$ the space of all formal power series $f(z, \zeta) = \sum_{v,w\in\mathcal{F}_d} f_{v,w} z^v \zeta^w$ in the two sets of indeterminants (z, ζ) with coefficients $f_{v,w} \in \mathcal{E}$. The subspace $\mathcal{P}(\mathcal{F}_d \times \mathcal{F}_d, \mathcal{E})$ of "polynomials" consists of all such formal series $f(z, \zeta) = \sum_{v,w\in\mathcal{F}_d} f_{v,w} z^v \zeta^w$ such that $f_{v,w} = 0$ for all but finitely many $v, w \in \mathcal{F}_d \times \mathcal{F}_d$. Given a block matrix $W = [W_{v,w;\alpha,\beta}]$ with rows and columns indexed by $\mathcal{F}_d \times \mathcal{F}_d$ as above, we may think of W as an operator from $\mathcal{P}(\mathcal{F}_d \times \mathcal{F}_d, \mathcal{E})$ to $L(\mathcal{F}_d, \times\mathcal{F}_d, \mathcal{E})$ by defining

$$f = Wp \text{ if } f_{v,w} = \sum_{\alpha,\beta} W_{v,w;\alpha,\beta} p_{\alpha,\beta} \text{ where } p(z,\zeta) = \sum_{\alpha,\beta} p_{\alpha,\beta} z^\alpha \zeta^\beta$$

(where the sum on the right is finite since p is assumed to be a polynomial). If W arises from a row-unitary as in (2.2.2), then W is positive semidefinite in the sense that

$$\langle Wp, p\rangle_{L^2} \geq 0 \text{ for all } p \in \mathcal{P}(\mathcal{F}_d \times \mathcal{F}_d, \mathcal{E}). \tag{2.2.7}$$

Here we are using that the formal L^2 inner product

$$\langle f, p\rangle_{L^2} = \sum_{v,w\in\mathcal{F}_d\times\mathcal{F}_d} \langle f_{v,w}, p_{v,w}\rangle_{\mathcal{E}}$$

gives a well-defined pairing between $L(\mathcal{F}_d \times \mathcal{F}_d, \mathcal{E})$ and $\mathcal{P}(\mathcal{F}_d \times \mathcal{F}_d, \mathcal{E})$, since the sum defining the inner product for such a pair collapses to a finite sum. We may therefore define a space $\mathcal{L}_W$ as the completion of $W\mathcal{P}(\mathcal{F}_d \times \mathcal{F}_d, \mathcal{E})$ in the inner product

$$\langle Wp, Wq\rangle_{\mathcal{L}_W} = \langle Wp, q\rangle_{L^2}.$$

Following the terminology of [**BaV04**], we say that W is a *Cuntz weight* if W is a positive semidefinite Haplitz operator with the Cuntz property, i.e., if W satisfies (2.2.3), (2.2.4), (2.2.5) and (2.2.7). So far we have argued that any W arising from a row-unitary $\mathcal{U}$ as in (2.2.2) is a Cuntz weight; for complete details, we refer to [**BaV04**].

Since, for any $f \in W\mathcal{P}(\mathcal{F}_d \times \mathcal{F}_d, \mathcal{E})$, $e \in \mathcal{E}$ and $v, w \in \mathcal{F}_d$ we have

$$\langle f, W[z^\alpha\zeta^\beta e]\rangle_{\mathcal{L}_W} = \langle f, z^\alpha\zeta^\beta e\rangle_{L^2} = \langle f_{\alpha\beta}, e\rangle_{\mathcal{E}}, \tag{2.2.8}$$

we see that the map $f \mapsto f_{v,w}$ of f to a given one of its Fourier coefficients is bounded in $\mathcal{L}_W$-norm, and hence elements of the completion f may still be identified as formal series $f(z,\zeta) = \sum_{v,w \in \mathcal{F}_d \times \mathcal{F}_d} f_{v,w} z^v \zeta^w$ in the space $\mathcal{L}_W$. It is natural to define a Fourier representation operator Φ by

$$\Phi\colon k \mapsto \sum_{v,w \in \mathcal{F}_d} (P_{\mathcal{E}} \mathcal{U}^w \mathcal{U}^{*v} k) z^v \zeta^w. \tag{2.2.9}$$

Then one can show that Φ is a coisometry mapping $\mathcal{K}$ onto $\mathcal{L}_W$. Moreover there is a model row unitary d-tuple $\mathcal{U}_W = (\mathcal{U}_{W,1} \dots, \mathcal{U}_{W,d})$ on $\mathcal{L}_W$ such that

$$\Phi \mathcal{U}_j = \mathcal{U}_{W,j} \Phi, \qquad \Phi \mathcal{U}_j^* = \mathcal{U}_{W,j}^* \Phi \text{ for } j = 1, \dots d, \tag{2.2.10}$$

and we recover W from Φ via

$$W = \Phi \Phi^{[*]} \text{ and } \mathcal{L}_W = \Phi \cdot \mathcal{K}. \tag{2.2.11}$$

The precise formula for $\mathcal{U}_W$ (at least on the dense subset $W\mathcal{P}(\mathcal{F}_d \times \mathcal{F}_d, \mathcal{E})$) is

$$\mathcal{U}_{W,j}\colon Wp \mapsto W S_j^R p \text{ for } p \in \mathcal{P}(\mathcal{F}_d \times \mathcal{F}_d, \mathcal{E}). \tag{2.2.12}$$

where S_j^R is the operator of multiplication by the indeterminate z_j on the right:

$$S_j^R\colon f(z,\zeta) \mapsto f(z,\zeta) \cdot z_j, \tag{2.2.13}$$

and can be viewed as a noncommutative version of a unilateral shift. The adjoint $\mathcal{U}_{W,j}^*$ of $\mathcal{U}_{W,j}$ is given densely by

$$\mathcal{U}_{W,j}^*\colon Wp \mapsto W U_j^{R[*]} p \tag{2.2.14}$$

where U_j^R is a noncommutative version of a bilateral shift:

$$U_j^R\colon f(z,\zeta) \mapsto f(0,\zeta) \cdot \zeta_j^{-1} + f(z,\zeta) \cdot z_j \tag{2.2.15}$$

where

$$f(0,\zeta) = \sum_{w \in \mathcal{F}_d} f_{\emptyset,w} \zeta^w \text{ if } f(z,\zeta) = \sum_{v,w \in \mathcal{F}_d} f_{v,w} z^v \zeta^w.$$

and $U_j^{R[*]}$ denotes the formal adjoint of U_j^R with respect to the L^2-inner product, given by

$$U_j^{R[*]}\colon f(z,\zeta) \mapsto f(0,\zeta) \cdot \zeta_j + f(z,\zeta) \cdot z_j^{-1}. \tag{2.2.16}$$

For the record, the formal adjoint $S_j^{R[*]}$ of S_j^R with respect to the L^2-inner product is given by

$$S_j^{R[*]}\colon f(z,\zeta) \mapsto f(z,\zeta) \cdot z_j^{-1}. \tag{2.2.17}$$

Here the same convention with respect to expressions of the form $z^v \cdot z^{w^{-1}}$ (and hence, in particular, for $z^v \cdot z_j^{-1} = z^v \cdot z^{g_j^{-1}}$) as was used for expressions of the form $\mathcal{U}^v \mathcal{U}^{w^{-1}}$ is in place (see (2.2.1) and the explanation there). In particular, if $\mathcal{E}$ is $*$-cyclic for the row-unitary $\mathcal{U} = (\mathcal{U}_1, \dots, \mathcal{U}_d)$, then Φ implements a unitary equivalence between the row-unitary $\mathcal{U}$ on $\mathcal{K}$ and the model row unitary $\mathcal{U}_W$ on $\mathcal{L}_W$. It turns out that the Haplitz property (2.2.3)–(2.2.4) imposed on a block-operator matrix W is equivalent to the intertwining condition

$$W S_j^R = U_j^R W \text{ on } \mathcal{P}(\mathcal{F}_d \times \mathcal{F}_d, \mathcal{E}). \tag{2.2.18}$$

Since $W = W^{[*]}$ (where $W^{[*]}$ is the adjoint of W with respect to the formal L^2-inner product pairing $\mathcal{P}(\mathcal{F}_d \times \mathcal{F}_d, \mathcal{E})$ with $L(\mathcal{F}_d \times \mathcal{F}_d, \mathcal{E})$), we see from (2.2.18) that we also have

$$WU_j^{R[*]} = S_j^{R[*]} W \text{ on } \mathcal{P}(\mathcal{F}_d \times \mathcal{F}_d, \mathcal{E}). \tag{2.2.19}$$

It then follows from (2.2.18) and (2.2.19) that $\mathcal{L}_W$ is invariant under U_j^R and $S_j^{R[*]}$ and that alternate formulas for $\mathcal{U}_{W,j}$ and $\mathcal{U}_{W,j}^*$ applicable for a general element of $\mathcal{L}_W$ are

$$\begin{aligned} \mathcal{U}_{W,j} f &= U_j^R f \\ \mathcal{U}_{W,j}^* f &= S_j^{R[*]} f \text{ for } f \in \mathcal{L}_W \end{aligned} \tag{2.2.20}$$

for $j = 1, \dots, d$. A consequence of the formulas (2.2.12) and (2.2.14) combined with the intertwining relations (2.2.18) and (2.2.19) are the following formulas for the action of $\mathcal{U}_{W,j}$ and $\mathcal{U}_{W,j}^*$ on a general element f of $\mathcal{L}_W$:

$$\mathcal{U}_{W,j} \colon f(z,\zeta) \mapsto f(0,\zeta) \cdot \zeta_j^{-1} + f(z,\zeta) \cdot z_j, \tag{2.2.21}$$

$$\mathcal{U}_{W,j}^* \colon f(z,\zeta) \mapsto f(0,\zeta) \cdot \zeta_j + f(z,\zeta) \cdot z_j^{-1}. \tag{2.2.22}$$

Conversely, suppose that W is any Cuntz weight. Since W is positive semidefinite (as in (2.2.7)), we may define a Hilbert space $\mathcal{L}_W$ as the completion of $W \cdot \mathcal{P}(\mathcal{F}_d \times \mathcal{F}_d, \mathcal{E})$ with respect to the inner product $\langle \cdot, \cdot \rangle_{\mathcal{L}_W}$ given by (2.1.1) and define operators $\mathcal{U}_{W,j}$ by (2.2.12) for $j = 1, \dots, d$. Then the Haplitz property of W guarantees that $\mathcal{U}_W = (\mathcal{U}_{W,1}, \dots, \mathcal{U}_{W,d})$ is row-isometric with the adjoint $\mathcal{U}_{W,j}^*$ of $\mathcal{U}_{W,j}$ given by (2.2.14) for $j = 1, \dots, d$. Furthermore, the Cuntz property (2.2.5) of W then guarantees that $\mathcal{U}_W$ is actually row-unitary. Moreover, the Fourier representation operator Φ_W given by (2.1.2) (with $\mathcal{U}_W$ in place of $\mathcal{U}$) is the identity on $\mathcal{L}_W$ and we have the concrete model version of (2.2.11):

$$W = \Phi_W \Phi_W^*, \qquad \mathcal{L}_W = \Phi_W \mathcal{L}_W. \tag{2.2.23}$$

It is often more convenient to work with the *symbol*

$$\widehat{W}(z,\zeta) = \sum_{v,w \in \mathcal{F}_d} \widehat{W}_{v,w} z^v \zeta^w$$

of W, the formal power series in (z,ζ) defined by

$$\widehat{W}(z,\zeta)e = (We)(z,\zeta) \text{ for } e \in \mathcal{E}$$

(where e is identified with the polynomial $e = ez^\emptyset \zeta^\emptyset \in \mathcal{P}(\mathcal{F}_d \times \mathcal{F}_d, \mathcal{E})$). As explained in [**BaV04**], any $[*]$-Haplitz operator (i.e., W for which both W and $W^{[*]}$ satisfy (2.2.3) and (2.2.4)) is uniquely determined by its symbol $\widehat{W}(z,\zeta)$ according to the formula

$$W_{v,w;\alpha,\beta} = \begin{cases} \widehat{W}_{(v\alpha^{-1})\beta^\top, w} & \text{if } |v| \geq |\alpha|, \\ \widehat{W}_{\beta^\top, w(\alpha v^{-1})^\top} & \text{if } |v| \leq |\alpha|, \end{cases} \tag{2.2.24}$$

and, conversely, any formal power series $\widehat{W}(z,\zeta) = \sum_{v,w \in \mathcal{F}_d} \widehat{W}_{v,w} z^v \zeta^w$ with coefficients $\widehat{W}_{v,w}$ equal to operators on $\mathcal{E}$ determines a $[*]$-Haplitz operator W by (2.2.24). The action of W on any polynomial $f(z,\zeta)$ can be computed directly from the symbol $\widehat{W}(z,\zeta)$ according to the formula

$$W[f](z,\zeta) = \widehat{W}(z',\zeta) k_{per}(z,\zeta) f(z, z'^{-1})|_{z'=0} + \widehat{W}(z',\zeta) f(z, z'^{-1})|_{z'=z}. \tag{2.2.25}$$

where we have set $k_{per}(z,\zeta)$ equal to the "perverse Szegö kernel"

$$k_{per}(z,\zeta) = \sum_{v' \neq \emptyset} (z^{-1})^{v'^\top} (\zeta^{-1})^{v'} \tag{2.2.26}$$

and where $z' = (z'_1, \dots, z'_d)$ is another set of noncommuting indeterminants, each of which commutes with $z_1, \dots, z_d, \zeta_1, \dots, \zeta_d$ (see Proposition 2.7 in [**BaV04**]). Here it is understood that one does the evaluation in z' followed by the evaluation $z' = z$ before performing the multiplication by $k_{per}(z,\zeta)$. For the case where $p(z) = \sum_\alpha p_\alpha z^\alpha$ is an analytic polynomial in $\mathcal{P}(\mathcal{F}_d \times \{\emptyset\}, \mathcal{E})$, then

$$W[p](z,\zeta) = \widehat{W}(0,\zeta) k_{per}(z,\zeta) p(z) + \widehat{W}(z,\zeta)p(z). \tag{2.2.27}$$

If in addition $\widehat{W}(0,\zeta) = I\zeta^\emptyset$, then (2.2.27) simplifies further to

$$W[p](z,\zeta) = \widehat{W}(z,\zeta)p(z) \text{ if } \widehat{W}(0,\zeta) = I\zeta^\emptyset. \tag{2.2.28}$$

The symbol of the adjoint $W^{[*]}$ of W is given by

$$\widehat{W^{[*]}}(z,\zeta) = \widehat{W}(\zeta, z)^*. \tag{2.2.29}$$

(where we use the convention that $(z^v)^* = z^{v^\top}$ and $(\zeta^w)^* = \zeta^{w^\top}$). Thus the selfadjointness of a $[*]$-Haplitz operator W can be expressed directly in terms of the symbol: $W = W^{[*]}$ *(as a $[*]$-Haplitz operator) if and only if*

$$\widehat{W}(z,\zeta) = \widehat{W}(\zeta, z)^* \tag{2.2.30}$$

Furthermore, a selfadjoint Haplitz operator is positive semidefinite if and only if its symbol $\widehat{W}(z,\zeta)$ and the *Cuntz defect* $D_{\widehat{W}}(z,\zeta)$ of its symbol given by

$$D_{\widehat{W}}(z,\zeta) = \widehat{W}(z,\zeta) - \sum_{k=1}^{d} z_k^{-1} \widehat{W}(z,\zeta)\zeta_k^{-1}.$$

have factorizations

$$\widehat{W}(z,\zeta) = Y(\zeta)Y(z)^*, \tag{2.2.31}$$

$$\widehat{W}(z,\zeta) - \sum_{j=1}^{d} z_j^{-1} \widehat{W}(z,\zeta)\zeta_j^{-1} = \Gamma(\zeta)\Gamma(z)^* \tag{2.2.32}$$

Here $Y(z)$ is a formal series of the form $Y(z) = \sum_{w \in \mathcal{F}_d} Y_w z^w$ where each Y_w is an operator from some auxiliary Hilbert space $\mathcal{H}$ to $\mathcal{E}$, and similarly, $\Gamma(z)$ is a formal series of the form $\Gamma(z) = \sum_{w \in \mathcal{F}_d} \Gamma_w z^w$. Furthermore, the Haplitz operator W is a Cuntz weight if and only if its symbol $\widehat{W}(z,\zeta)$ is a positive symbol (i.e., a factorization (2.2.31) exists) and its Cuntz defect $D_{\widehat{W}}(z,\zeta)$ is zero:

$$\widehat{W}(z,\zeta) - \sum_{j=1}^{d} z_j^{-1} \widehat{W}(z,\zeta)\zeta_j^{-1} = 0. \tag{2.2.33}$$

Moreover, the Haplitz operator $W = [W_{v,w;\alpha,\beta}]$ with matrix entries

$$W_{v,w;\alpha,\beta} = P_\mathcal{E} \mathcal{U}^w \mathcal{U}^{*v} \mathcal{U}^{\alpha^\top} \mathcal{U}^{*\beta^\top}|_\mathcal{E} \tag{2.2.34}$$

(or, equivalently, its symbol $\widehat{W}(z,\zeta)$) is a *complete* unitary invariant for a row-unitary $\mathcal{U}$ together with a specified choice of $*$-cyclic subspace $\mathcal{E}$. For complete details, we refer to [**BaV04**].

2.3. Hardy subspaces

An analogue of the Hardy space inside $\mathcal{L}_W$ is the space $\mathcal{H}_W$ defined by

$$\mathcal{H}_W := \text{ closure in } \mathcal{L}_W \text{ of } W\mathcal{P}(\mathcal{F}_d \times \{\emptyset\}, \mathcal{E}).$$

The subspace $\mathcal{H}_W$ is invariant under $\mathcal{U}_{W,j}$ for each $j = 1, \dots, d$, since

$$(\mathcal{U}_{W,j} W)[p](z) = (WS_j^R)[p](z) = W[p(z) \cdot z_j] \in \mathcal{H}_W.$$

Hence the restriction $\mathcal{U}_W^+ := \mathcal{U}_W|_{\mathcal{H}_W}$ of $\mathcal{U}_W$ to $\mathcal{H}_W$ is a row isometry. For $\mathcal{U}_W^+$ to be in fact row unitary, it must be the case that $\mathcal{U}_{W,1}\mathcal{H}_W + \dots + \mathcal{U}_{W,d}\mathcal{H}_W$ is equal to $\mathcal{H}_W$. This in turn implies that $\mathcal{H}_W$ is invariant for each $\mathcal{U}_{W,j}^*$ from which it follows that $\mathcal{H}_W = \mathcal{L}_W$.

Note that

$$\langle Wp, Wq \rangle_{\mathcal{H}_W} = \langle Wp, q \rangle_{L^2(\mathcal{F}_d \times \mathcal{F}_d, \mathcal{E})} = \langle W^+ p, q \rangle_{L^2(\mathcal{F}_d, \mathcal{E})} \tag{2.3.1}$$

where W^+ is the $\mathcal{F}_d \times \mathcal{F}_d$ block matrix

$$W^+ = [W^+_{v,\alpha}]_{v,\alpha \in \mathcal{F}_d} \text{ where } W^+_{v,\alpha} = W_{v,\emptyset;\alpha,\emptyset}. \tag{2.3.2}$$

Given any positive semidefinite $\mathcal{F}_d \times \mathcal{F}_d$ matrix $V = [V_{v,\alpha}]_{v,\alpha \in \mathcal{F}_d}$ (with matrix entries equal to bounded operators on the Hilbert space $\mathcal{E}$), let us define a related analogue of Hardy space $\mathcal{H}_V^+$ by $\mathcal{H}_V^+ = \text{closure}\, V\mathcal{P}(\mathcal{F}_d, \mathcal{E})$ in the inner product

$$\langle Vp, Vq \rangle_{\mathcal{H}_V^+} = \langle Vp, q \rangle_{L^2(\mathcal{F}_d, \mathcal{E})}. \tag{2.3.3}$$

In case $V = W^+$ with the $\mathcal{F}_d \times \mathcal{F}_d$ matrix W^+ coming from a positive semidefinite $(\mathcal{F}_d \times \mathcal{F}_d) \times (\mathcal{F}_d \times \mathcal{F}_d)$ matrix as in (2.3.2), the import of the calculation (2.3.1) is that the map

$$f(z, \zeta) = \sum_{v,w} f_{v,w} z^v \zeta^w \mapsto f(z, 0) := \sum_v f_{v,\emptyset} z^v$$

is a unitary transformation from $\mathcal{H}_W$ onto $\mathcal{H}_{W^+}^+$. It will often be convenient to work with the spaces $\mathcal{H}_{W^+}^+$ (which sit inside $L(\mathcal{F}_d, \mathcal{E})$ and depend only on W^+ rather than the full W) rather than $\mathcal{H}_W$ when analyzing issues not depending on the embedding of $\mathcal{H}_W$ inside $\mathcal{L}_W$.

There is a model theory for row isometries with given cyclic subspace $\mathcal{E}$ which is parallel to (and simpler than) that sketched above for row unitaries. If $\mathcal{S} = (\mathcal{S}_1, \dots, \mathcal{S}_d)$ is a row isometry on the Hilbert space $\mathcal{H}$ with cyclic subspace $\mathcal{E}$, we form the $\mathcal{F}_d \times \mathcal{F}_d$ matrix

$$W^+ = [W^+_{v,\alpha}]_{v,\alpha \in \mathcal{F}_d} \text{ with } W^+_{v,\alpha} = P_{\mathcal{E}} \mathcal{S}^{*v} \mathcal{S}^{\alpha^\top}|_{\mathcal{E}}. \tag{2.3.4}$$

Then W^+ has the Toeplitz property

$$W^+_{v,\alpha g_j} = W^+_{v g_j^{-1},\alpha} \text{ for } v, \alpha \in \mathcal{F}_d \text{ and } j = 1, \dots d, \tag{2.3.5}$$

and $\mathcal{S}$ is unitarily equivalent to $\mathcal{S}_{W^+} = (\mathcal{S}_{W^+,1}, \dots, \mathcal{S}_{W^+,d})$ on $\mathcal{H}_{W^+}^+$ given by

$$\mathcal{S}_{W^+,j} \colon W^+ p \mapsto W^+ S_j^R p \text{ for } p \in \mathcal{P}(\mathcal{F}_d, \mathcal{E}). \tag{2.3.6}$$

Conversely, if W^+ is a positive semidefinite $\mathcal{F}_d \times \mathcal{F}_d$ operator matrix enjoying the Toeplitz property (2.3.5), then $\mathcal{S}_{W^+}$ defined by (2.3.6) is a row isometry on $\mathcal{H}_{W^+}^+$.

In case W^+ is a $\mathcal{F}_d \times \mathcal{F}_d$ matrix having the Toeplitz property

$$W^+_{v,\alpha g_j} = W^+_{v g_j^{-1},\alpha} \text{ for } v, \alpha \in \mathcal{F}_d \text{ and } j = 1, \dots d,$$

then the d-tuple of operators $\mathcal{S}_{W^+} = (\mathcal{S}_{W^+,1}, \dots, \mathcal{S}_{W^+,d})$ given by

$$S_{W^+,j} \colon W[f](z) \mapsto\colon W[f(z)z_j]$$

is a row-isometry with cyclic subspace $W^+\mathcal{E} \subset \mathcal{H}_{W^+}$ and the positive Cuntz-Toeplitz matrix W^+ (2.3.4) is a complete unitary invariant for the row isometry $\mathcal{S}$ with given cyclic subspace $\mathcal{E}$.

In case W^+ is the identity operator ($W^+_{v,\alpha} = \delta_{v,\alpha} I_{\mathcal{E}}$), then $\mathcal{H}^+_{W^+} = L^2(\mathcal{F}_d, \mathcal{E})$ isometrically, the subspace $W^+\mathcal{E}$ is *wandering* for $\mathcal{S} = \mathcal{S}_{W^+}$ (i.e. $\mathcal{S}^v\mathcal{E} \perp \mathcal{S}^{v'}\mathcal{E}$ for all $v \neq v'$ in $\mathcal{F}_d$) and $\mathcal{S}$ is a *row shift*, i.e., a row isometry on a Hilbert space $\mathcal{H}$ with the additional property that

$$\bigcap_{N=0}^{\infty} \operatorname{closed\,span}_{v \in \mathcal{F}_d \colon |v|=N} \mathcal{S}^v\mathcal{H} = \{0\}. \tag{2.3.7}$$

For a given positive semidefinite Haplitz W, it will be useful to understand the orthogonal projection $P_{\mathcal{H}_W}$ from $\mathcal{L}_W$ to $\mathcal{H}_W$. The following proposition will be useful.

PROPOSITION 2.3.1. *An element $f(z,\zeta) = \sum_{v,w \in \mathcal{F}_d} f_{v,w} z^v \zeta^w$ in $\mathcal{L}_W$ and its projection $P_{\mathcal{H}_W} f(z,\zeta) = \sum_{v,w \in \mathcal{F}_d} (P_{\mathcal{H}_W} f)_{v,w} z^v \zeta^w$ to $\mathcal{H}_W$ have the same Fourier coefficients on the future $\mathcal{F}_d \times \{\emptyset\}$:*

$$f_{v,\emptyset} = (P_{\mathcal{H}_W} f)_{v,\emptyset} \text{ for all } v \in \mathcal{F}_d. \tag{2.3.8}$$

The orthogonal complement $\mathcal{H}_W^\perp$ of $\mathcal{H}_W$ in $\mathcal{L}_W$ is characterized as

$$\mathcal{H}_W^\perp = \{f = \sum_{v,w \in \mathcal{F}_d} f_{v,w} z^v \zeta^w \in \mathcal{L}_W \colon f_{v,\emptyset} = 0 \text{ for all } v \in \mathcal{F}_d\}. \tag{2.3.9}$$

PROOF. By (2.2.8) we that $\langle f_{v,w}, e\rangle_{\mathcal{E}} = \langle f, W[ez^v\zeta^w]\rangle_{\mathcal{L}_W}$ for $f \in \mathcal{L}_W$ and $e \in \mathcal{E}$. In case $w = \emptyset$, we have $W[ez^v\zeta^\emptyset] \in \mathcal{H}_W$ and hence

$$\begin{aligned}
\langle f_{v,\emptyset}, e\rangle_{\mathcal{E}} &= \langle f(z,\zeta), W[ez^v\zeta^\emptyset]\rangle_{\mathcal{L}_W} \\
&= \langle (P_{\mathcal{H}_W} f)(z,\zeta), W[ez^v\zeta^\emptyset]\rangle_{\mathcal{L}_W} \\
&= \langle (P_{\mathcal{H}_W} f)_{v,\emptyset}, e\rangle_{\mathcal{E}}
\end{aligned}$$

and the first assertion of Proposition 2.3.1 follows.

From the definition of $\mathcal{H}_W$ we see that $\mathcal{H}_W^\perp$ is characterized as the orthogonal complement in $\mathcal{L}_W$ of $W\mathcal{P}(\mathcal{F}_d \times \{\emptyset\}, \mathcal{E})$. Since in general we have

$$\langle f, W[ez^v\zeta^\emptyset]\rangle_{\mathcal{L}_W} = \langle f_{v,\emptyset}, e\rangle_{\mathcal{E}}$$

we see that equivalently $\mathcal{H}_W^\perp$ is characterized as the set of all $f \in \mathcal{L}_W$ with $f_{v,\emptyset} = 0$ for all $v \in \mathcal{F}_d$, and the rest of the proposition follows. □

REMARK 2.3.2. We emphasize that it is not the case that $f_{v,w} = 0$ for $w \neq \emptyset$ and $f(z,\zeta) = \sum_{v,w \in \mathcal{F}_d} f_{v,w} z^v \zeta^w \in \mathcal{H}_W$. Indeed, the symbol of W applied to a vector $e \in \mathcal{E}$ given by

$$\widehat{W}(z,\zeta)e = W[ez^\emptyset\zeta^\emptyset](z,\zeta)$$

is in $\mathcal{H}_W$ with $f_{v,w} = W_{v,w;\emptyset,\emptyset}e = W_{\emptyset,w;\emptyset,v^\top}e$ (by (2.2.6)). As we must have

$$\sum_{j=1}^{d} W_{\emptyset,g_j;\emptyset,g_j} = W_{\emptyset,\emptyset;\emptyset,\emptyset} = W^+_{\emptyset,\emptyset}$$

by (2.2.5), it is certainly atypical for matrix entries of the form $W_{\emptyset,w;\emptyset,v^\top}$ to vanish.

A particular class of $[*]$-Haplitz operators $W = [W_{v,w;\alpha,\beta}]_{v,w,\alpha,\beta\in\mathcal{F}_d}$ are those for which $W_{v,\emptyset;\alpha,\emptyset} = \delta_{v,\alpha} I_{\mathcal{E}}$; then we shall say that W *is a Haplitz extension of the identity*. In this case, from (2.2.3) we see that

$$\widehat{W}_{v,w} = W_{v,w;\emptyset,\emptyset} = W_{v,\emptyset;w^\top,\emptyset} = \delta_{v,w} I_{\mathcal{E}}.$$

In particular $\widehat{W}_{\emptyset,w} = \delta_{\emptyset,w} I_{\mathcal{E}}$ and $W_{v,\emptyset} = \delta_{v,\emptyset} I_{\mathcal{E}}$ from which we get $\widehat{W}(z,0) = I_{\mathcal{E}} z^{\emptyset}$ and $\widehat{W}(0,\zeta) = I_{\mathcal{E}}\zeta^{\emptyset}$. Hence we are in the situation where formula (2.2.28) applies for the action of $\widehat{W}$ on an analytic polynomial. We have verified the first part of the following proposition concerning positive-semidefinite Haplitz extensions of the identity.

PROPOSITION 2.3.3. *Suppose that the positive semidefinite Haplitz operator W is a Haplitz extension of the identity and that $f \in \mathcal{L}_W$.*

(1) *Then the action of W on an an element $f \in L^2(\mathcal{F}_d \times \{\emptyset\}, \mathcal{E})$ is given by $W[f](z,\zeta) = \widehat{W}(z,\zeta) f(z)$.*
(2) *Then the orthogonal projection $P_{\mathcal{H}_W} f$ of f onto $\mathcal{H}_W$ is given by*

$$P_{\mathcal{H}_W} f = W\left[\sum_{v\in\mathcal{F}_d} f_{v,\emptyset} z^v\right]. \tag{2.3.10}$$

PROOF. The first statement was verified in the discussion preceding the statement of the proposition, so it suffices to consider only the second statement.

Set $f' = W\left[\sum_{v\in\mathcal{F}_d} f_{v,\emptyset} z^v\right]$. Since W is a Haplitz extension of the identity, it is clear that $\sum_{v\in\mathcal{F}_d} f_{v,\emptyset} z^v$ is in $L^2(\mathcal{F}_d,\mathcal{E})$ and that $f' \in \mathcal{H}_W$. Again since W is a Haplitz extension of the identity, we see that $f_{v,\emptyset} = f'_{v,\emptyset}$ for all $v \in \mathcal{F}_d$. It now follows from Proposition 2.3.1 that $f' = P_{\mathcal{H}_W} f$ as wanted. □

In case it happens that $\mathcal{H}_W \neq \mathcal{L}_W$, an interesting issue is the classification of all row unitary extensions $(\mathcal{U}_1,\dots,\mathcal{U}_d)$ of the given row isometry $\mathcal{U}_W^+$. More abstractly, we can ask for all *minimal* row-unitary extensions $\mathcal{U} = (\mathcal{U}_1,\dots,\mathcal{U}_d)$ of a given row isometry $\mathcal{S} = (\mathcal{S}_1,\dots,\mathcal{S}_d)$ on $\mathcal{H}$, i.e., a row unitary $\mathcal{U} = (\mathcal{U}_1,\dots,\mathcal{U}_d)$ on a Hilbert space $\mathcal{K} \supset \mathcal{H}$ such that $S_j = \mathcal{U}_j|_{\mathcal{H}}$ and $\mathcal{K} = \text{closed span}_{v,w\in\mathcal{F}_d} \mathcal{U}^w\mathcal{U}^{*v}\mathcal{H}$. As laid out in the next Proposition, an equivalent problem is the classification of all Cuntz-weight extensions of a given positive semidefinite Toeplitz weight.

PROPOSITION 2.3.4. *Let $\mathcal{S} = (\mathcal{S}_1,\dots,\mathcal{S}_d)$ be a row isometry on a Hilbert space $\mathcal{H}$ with cyclic subspace $\mathcal{E}$, and define a positive semidefinite Toeplitz operator $W^+ = [W^+]_{v,\alpha\in\mathcal{F}_d}$ by $W^+_{v,\alpha} = P_{\mathcal{E}}\mathcal{S}^{*v}\mathcal{S}^{\alpha^\top}|_{\mathcal{E}}$. Then minimal row unitary extensions $\mathcal{U} = (\mathcal{U}_1,\dots,\mathcal{U}_d)$ of $\mathcal{S}$ are in one-to-one correspondence with Cuntz-weight extensions $W = [W_{w,v;\alpha,\beta}]_{w,v,\alpha,\beta\in\mathcal{F}_d}$ of the Cuntz-Toeplitz operator W^+, i.e., Cuntz weights W on $\mathcal{E}$ such that*

$$W_{v,\emptyset;\alpha,\emptyset} = W^+_{v,\alpha}.$$

PROOF. If $\mathcal{U} = (\mathcal{U}_1,\dots,\mathcal{U}_d)$ is a row unitary extension of $\mathcal{S}$ on $\mathcal{K} \supset \mathcal{H}$, set

$$W_{v,w;\alpha,\beta} = P_{\mathcal{E}}\mathcal{U}^w\mathcal{U}^{*v}\mathcal{U}^{\alpha^\top}\mathcal{U}^{*\beta^\top}|_{\mathcal{E}}. \tag{2.3.11}$$

Then we see that $W_{v,\emptyset;\alpha,\emptyset} = P_{\mathcal{E}}\mathcal{U}^{*v}\mathcal{U}^{\alpha^\top}|_{\mathcal{E}} = P_{\mathcal{E}}\mathcal{S}^{*v}\mathcal{S}^{\alpha^\top}|_{\mathcal{E}} = W^+_{v,\alpha}$.

Conversely, let $\mathcal{U}$ be a minimal row-unitary extension of the row-isometry $\mathcal{S}$ which has cyclic subspace $\mathcal{E}$. Then $\mathcal{S}$ is unitarily equivalent to $\mathcal{S}_{W^+}$ on $\mathcal{H}_{W^+}$

where $W^{+}_{v,\alpha} = P_{\mathcal{E}}\mathcal{S}^{*v}\mathcal{S}^{\alpha^{\top}}|_{\mathcal{E}}$ while $\mathcal{U}$ is unitarily equivalent to $\mathcal{U}_W$ on $\mathcal{L}_W$ where $W_{v,w;\alpha,\beta} = P_{\mathcal{E}}\mathcal{U}^{w}\mathcal{U}^{*v}\mathcal{U}^{\alpha^{\top}}\mathcal{U}^{*\beta^{\top}}|_{\mathcal{E}}$. Thus, for $e, e' \in \mathcal{E}$, we have

$$\begin{aligned} \langle W_{v,\emptyset;\alpha,\emptyset}e, e'\rangle_{\mathcal{E}} &= \langle \mathcal{U}^{\alpha^{\top}} e, \mathcal{U}^{v^{\top}} e'\rangle_{\mathcal{K}} \\ &= \langle \mathcal{S}^{\alpha^{\top}} e, \mathcal{S}^{v^{\top}} e'\rangle_{\mathcal{H}} \\ &= \langle P_{\mathcal{E}}\mathcal{S}^{*v}\mathcal{S}^{\alpha^{\top}} e, e'\rangle_{\mathcal{E}} \\ &= \langle W^{+}_{v,\alpha}e, e'\rangle_{\mathcal{E}} \end{aligned}$$

and we see that W is a Cuntz-weight extension of W^{+}. Conversely, if W is a Haplitz extension of W^{+}, then $\mathcal{U}_W$ is a row unitary extension of $\mathcal{S}_{W^{+}}$. This correspondence between row unitary extensions $\mathcal{U}$ of $\mathcal{S}$ and Cuntz-weight extensions W of W^{+} is injective (up to unitary equivalence of row unitary extensions fixing the subspace $\mathcal{H}$ on which $\mathcal{S}$ acts) by the result from [**BaV04**] that the Cuntz weight (2.2.34) is a complete unitary invariant for a row unitary $\mathcal{U}$ with specified $*$-cyclic subspace $\mathcal{E}$. □

By the symbol calculus, the problem of extending a positive-definite Toeplitz weight W^{+} to a Cuntz weight W can be expressed in terms of the symbol $\widehat{W}(z,\zeta)$ of W: *given a collection* $W^{+}_{v,\emptyset}$ *(for* $v \in \mathcal{F}_d$*) with* $W^{+}_{\emptyset,\emptyset} = (W^{+}_{\emptyset,\emptyset})^{*}$, *construct a formal power series* $\widehat{W}(z,\zeta)$ *satisfying conditions* (2.2.31) *and* (2.2.33) *which in addition satisfies the boundary condition*

$$\widehat{W}(0,\zeta) = \sum_{w\in\mathcal{F}_d} (W^{+}_{w^{\top},\emptyset})^{*}\zeta^{w}. \tag{2.3.12}$$

The following proposition lays out the algebraic freedom in this extension process.

PROPOSITION 2.3.5. *Suppose that* $W^{+}_{v,\emptyset}$ *is a given collection of operators on* $\mathcal{E}$ *with* $W^{+}_{\emptyset,\emptyset} = (W^{+}_{\emptyset,\emptyset})^{*}$. *Then*

$$\widehat{W}(z,\zeta) = \sum_{v,w\in\mathcal{F}_d} \widehat{W}_{v,w} z^{v}\zeta^{w}$$

is a formal power series satisfying the conditions (2.2.30), (2.2.33) *along with the boundary condition* (2.3.12), *namely*

$$\begin{aligned} \widehat{W}(z,\zeta) &= \widehat{W}(z,\zeta)^{*}, \\ \widehat{W}(z,\zeta) &= \sum_{j=1}^{d} z_j^{-1}\widehat{W}(z,\zeta)\zeta_j^{-1} \\ \widehat{W}(0,\zeta) &= \sum_{w\in\mathcal{F}_d} (W^{+}_{w^{\top},\emptyset})^{*}\zeta^{w}, \end{aligned}$$

if and only if the coefficients $\widehat{W}_{v,w}$ *satisfy the following inductively nested system of linear equations:*

(1) *Fix* $k \in \{0,1,2,\dots\}$ *and inductively assume that* $\widehat{W}_{v,w}$ *has been constructed for* $|v| = |w|+k$ *and* $|w| = N$. *Construct* $\widehat{W}_{v',w'}$ *for* $|v'| = |w'|+k$ *and* $|w'| = N+1$ *subject to the linear equations*

$$\sum_{j=1}^{d} \widehat{W}_{g_j v, w g_j} = \widehat{W}_{v,w}$$

for all $v, w \in \mathcal{F}_d$ *with* $|v| = |w| + k$ *and* $|w| = N$, *subject to the initial condition*

$$\widehat{W}_{v,\emptyset} = W^+_{v,\emptyset} \text{ for } |v| = k.$$

If $k = 0$, *arrange also that* $\widehat{W}_{v',w'} = (\widehat{W}_{w'^\top, v'^\top})^*$.

(2) *Apply the construction of Step 1 to obtain a choice of* $\widehat{W}_{v,w}$ *for all* $v, w \in \mathcal{F}_d$ *with* $|v| \geq |w|$. *For* $|w| \geq |v|$, *set* $\widehat{W}_{v,w} = (\widehat{W}_{w^\top, v^\top})^*$.

REMARK 2.3.6. Note that Proposition 2.3.5 arranges only the selfadjointness condition $\widehat{W}(z,\zeta) = \widehat{W}(z,\zeta)^*$ rather than the stronger positivity condition

$$W(z,\zeta) = Y(\zeta)Y(z)^*$$

as in (2.2.31). This latter condition must be analyzed separately. In the next section we shall give a structural result which reduces the problem for a general positive semidefinite Toeplitz weight W^+ to the special case where W^+ is the identity operator ($W^+_{v,\alpha} = \delta_{v,\alpha} I_{\mathcal{E}}$).

2.4. Intertwining operators

A noncommutative analogue of Hardy space much studied of late is the space $L^2(\mathcal{F}_d, \mathcal{E})$ of formal power series $f(z) = \sum_{v \in \mathcal{F}_d} f_v z^v$ with $\|f\|^2_{L^2} = \sum_{v \in \mathcal{F}_d} \|f_v\|^2 < \infty$. The d-tuple $\mathcal{S}_{\mathcal{E}} = (\mathcal{S}_{\mathcal{E},1} \dots, \mathcal{S}_{\mathcal{E},d})$ of right shift operators acting on $L^2(\mathcal{F}_d, \mathcal{E})$

$$\mathcal{S}^R_{\mathcal{E},j} \colon f(z) \mapsto f(z) z_j \text{ for } j = 1, \dots, d$$

is a row shift (see (2.3.7)), and in fact is the model for any row shift of multiplicity equal to $\dim \mathcal{E}$. If an operator $M \colon L^2(\mathcal{F}_d, \mathcal{E}) \to L^2(\mathcal{F}_d, \mathcal{E}_*)$ intertwines $\mathcal{S}_{\mathcal{E}}$ with $\mathcal{S}_{\mathcal{E}_*}$

$$M \mathcal{S}_{\mathcal{E},j} = \mathcal{S}_{\mathcal{E}_*,j} M \text{ for } j = 1, \dots, d$$

then it is not difficult to show that there is a formal power series $T(z) = \sum_{v \in \mathcal{F}_d} T_v z^v$, with coefficients T_v equal to bounded operators from $\mathcal{E}$ to $\mathcal{E}_*$, such that $M = M_T$ where M_T is the multiplication operator

$$M_T \colon p(z) \mapsto T(z) \cdot p(z)$$

defined initially only on polynomials $\mathcal{P}(\mathcal{F}_d, \mathcal{E})$ and then extended to the whole space. The set of multipliers $T(z)$ inducing bounded multiplication operators $M_T \colon L^2(\mathcal{F}_d, \mathcal{E}) \to L^2(\mathcal{F}_d, \mathcal{E}_*)$ in this way then is a noncommutative analogue of operator-valued H^∞. We shall be particularly interested in the class of such $T(z)$ for which $\|M_T\| \leq 1$. The set of all such $T(z)$ then is a noncommutative analogue of the Schur-class of analytic functions mapping the unit disk into the closed unit disk. We shall use the notation $\mathcal{S}_{nc,d}(\mathcal{E}, \mathcal{E}_*)$ to denote this *noncommutative, d-variable Schur class*:

(2.4.1)
$$\mathcal{S}_{nc,d}(\mathcal{E}, \mathcal{E}_*) = \{T(z) = \sum_{v \in \mathcal{F}_d} T_v z^v \colon M_T \colon L^2(\mathcal{F}_d, \mathcal{E}) \to L^2(\mathcal{F}_d, \mathcal{E}_*) \text{ has } \|M_T\| \leq 1\}.$$

Our main concern here is a noncommutative analogue of the class of Laurent operators intertwining two bilateral shifts. Given two Cuntz weights W and W_* (with matrix entries equal to operators on Hilbert spaces $\mathcal{E}$ and $\mathcal{E}_*$ respectively), we say that the operator $S \colon \mathcal{L}_W \to \mathcal{L}_{W_*}$ is an *intertwining operator* if

$$S \mathcal{U}_{W,j} = \mathcal{U}_{W_*,j} S, \qquad S \mathcal{U}^*_{W,j} = \mathcal{U}^*_{W_*,j} S \tag{2.4.2}$$

for $j = 1, \dots, d$. If in addition S preserves the respective Hardy spaces ($S\colon \mathcal{H}_W \to \mathcal{H}_{W_*}$), we say that S is an *analytic intertwining operator*. Of particular interest is the class $\mathcal{S}(W, W_*)$ (the *noncommutative Schur multiplier class* associated with Cuntz weights W and W_*) of contractive, analytic intertwining operators between $\mathcal{L}_W$ and $\mathcal{L}_{W_*}$.

By definition, an analytic intertwining operator S maps $\mathcal{H}_W$ into $\mathcal{H}_{W_*}$. A nice subclass of such operators consists of those S which map the dense subset $W\mathcal{P}(\mathcal{F}_d \times \{\emptyset\}, \mathcal{E})$ of $\mathcal{H}_W$ into the dense subset $W_*\mathcal{P}(\mathcal{F}_d \times \{\emptyset\}, \mathcal{E}_*)$ of $\mathcal{H}_{W_*}$. In this case there is a formal power series $T(z) = \sum_{v \in \mathcal{F}_d} T_v z^v$ with coefficients equal to bounded operators from $\mathcal{E}$ to $\mathcal{E}_*$ such that

$$S\colon W[p] \to W_*[M_T p]$$

where M_T is the operator defined on the space of all analytic power series $L(\mathcal{F}_d \times \{\emptyset\}, \mathcal{E})$ by

$$(M_T f)(z) = T(z) \cdot f(z) := \sum_{w \in \mathcal{F}_d} \left(\sum_{v, v' \colon vv' = w} T_v f_{v'} \right) z^w \text{ if } f(z) = \sum_{w \in \mathcal{F}_d} f_w z^w.$$

We may extend M_T to an operator which we call L_T defined on $L_{\text{fin}-}(\mathcal{F}_d \times \mathcal{F}_d, \mathcal{E})$ (the space of formal power series $f(z, \zeta) = \sum_{v, w \in \mathcal{F}_d} f_{v,w} z^v \zeta^w$ where $f_{v,w} = 0$ for all but finitely many $(v, w) \in \mathcal{F}_d \times \mathcal{F}_d$ with $w \neq \emptyset$) by demanding

$$L_T[z^v \zeta^w e] = S^{Rv^\top} U^{R[*]w^\top} [T(z)e] \text{ for } e \in \mathcal{E},\ v, w \in \mathcal{F}_d$$

and extending by linearity to $\mathcal{P}(\mathcal{F}_d \times \mathcal{F}_d, \mathcal{E})$. Then

$$L_T|_{\mathcal{P}(\mathcal{F}_d \times \{\emptyset\}, \mathcal{E})} = M_T|_{\mathcal{P}(\mathcal{F}_d \times \{\emptyset\}, \mathcal{E})}.$$

The defining properties of $L_T\colon \mathcal{P}(\mathcal{F}_d \times \mathcal{F}_d, \mathcal{E}) \to L(\mathcal{F}_d \times \mathcal{F}_d, \mathcal{E})$ are:

$$L_T|_{\mathcal{P}(\mathcal{F}_d \times \{\emptyset\}, \mathcal{E})} = M_T|_{\mathcal{P}(\mathcal{F}_d \times \{\emptyset\}, \mathcal{E})}, \tag{2.4.3}$$

$$L_T U_j^{R[*]} = U_j^{R[*]} L_T \text{ for } j = 1, \dots, d \text{ on } \mathcal{P}(\mathcal{F}_d \times \mathcal{F}_d, \mathcal{E}) \tag{2.4.4}$$

$$L_T S_j^R = S_j^R L_T \text{ for } j = 1, \dots, d \text{ on } \mathcal{P}(\mathcal{F}_d \times \mathcal{F}_d, \mathcal{E}). \tag{2.4.5}$$

Viewing $\mathcal{L}(\mathcal{F}_d \times \mathcal{F}_d, \mathcal{E})$ as the dual of $\mathcal{P}(\mathcal{F}_d \times \mathcal{F}_d, \mathcal{E})$ in the L^2-inner product, we see that $L_T^{[*]}$ is well defined as an operator of the form $L_T^{[*]}\colon \mathcal{P}(\mathcal{F}_d \times \mathcal{F}_d, \mathcal{E}_*) \to \mathcal{L}(\mathcal{F}_d \times \mathcal{F}_d, \mathcal{E})$.

Formulas for the action of L_T and its L^2-adjoint $L_T^{[*]}$ obtained in [**BaV04**, Proposition 3.1] are

$$L_T[f](z, \zeta) = T(\zeta^{-1}) f(z, \zeta) - [T(\zeta^{-1}) f(z, \zeta)]|_{\zeta = 0} + T(z') f(z, z'^{-1})|_{z' = z}, \tag{2.4.6}$$

$$L_T^{[*]}[g](z, \zeta) = T(\zeta)^* \left[g(z, \zeta) + k_{per}(z, \zeta) g(z, 0) \right] \tag{2.4.7}$$

where we use the convention $T(\zeta)^* = \sum_{v \in \mathcal{F}_d} T_v^* \zeta^{v^\top}$, where $k_{per}(z, \zeta)$ is the "perverse Szegö kernel" as in (2.2.26), and where it is understood that the z'-variables are placed to the left of the z-variables in the last term of (2.4.6) before the evaluation $z' = z$. From these formulas we see that L_T and $L_T^{[*]}$ are actually well defined on all of $L(\mathcal{F}_d \times \mathcal{F}_d, \mathcal{E})$ and $L(\mathcal{F}_d \times \mathcal{F}_d, \mathcal{E}_*)$; this is analogous to the fact that the operator $M_T\colon f(z) \mapsto T(z) f(z)$ is well-defined as an operator on formal power series $M_T\colon L(\mathcal{F}_d, \mathcal{E}) \to L(\mathcal{F}_d, \mathcal{E}_*)$, as only finite sums are involved in the calculation of a given coefficient of $T(z) \cdot f(z)$ in terms of the (infinitely many) coefficients of $f(z)$.

Under the assumption that $L_T p$ is in the domain of the Haplitz operator W_*, the intertwining conditions (2.4.2) then determine the action of $S = L_T^{W,W_*}$ on all of $W\mathcal{P}(\mathcal{F}_d \times \mathcal{F}_d, \mathcal{E})$ according to the formula

$$L_T^{W,W_*}: Wp \mapsto W_* L_T p. \tag{2.4.8}$$

The setting discussed in [**BaV04**] is the case where $T(z) = \sum_v T_v z^v$ is assumed to be a polynomial (so $T_v = 0$ for all but finitely many T_v). Then L_T maps polynomials $\mathcal{P}(\mathcal{F}_d \times \mathcal{F}_d, \mathcal{E})$ into polynomials $\mathcal{P}(\mathcal{F}_d \times \mathcal{F}_d, \mathcal{E}_*)$ which is always contained in the domain of a Haplitz operator W_*. Hence $L_T^{W,W_*}: W\mathcal{P}(\mathcal{F}_d \times \mathcal{F}_d, \mathcal{E}) \to W_*\mathcal{P}(\mathcal{F}_d \times \mathcal{F}_d, \mathcal{E}_*)$ is well-defined, and $W_* L_T$, $L_T^{[*]}$ and $L_T^{[*]} W_* L_T$ are all well-defined Haplitz operators.

There is another setting where one can define the action of L_T^{W,W_*} on a domain larger than the polynomials. Specifically, suppose that W and W_* are Haplitz extensions of the identity and M_T defines a bounded operator from $L^2(\mathcal{F}_d, \mathcal{E})$ into $L^2(\mathcal{F}_d, \mathcal{E}_*)$. Since W_* is a Haplitz extension of the identity, then the domain of W_* includes the space

$$L^2_{\text{fin}-}(\mathcal{F}_d \times \mathcal{F}_d, \mathcal{E}) = \mathcal{P}(\mathcal{F}_d \times (\mathcal{F}_d \setminus \{\emptyset\}), \mathcal{E}) + L^2(\mathcal{F}_d \times \{\emptyset\}, \mathcal{E}).$$

Moreover, since M_T maps $L^2(\mathcal{F}_d, \mathcal{E})$ into $L^2(\mathcal{F}_d, \mathcal{E}_*)$, it follows that L_T maps $L^2_{\text{fin}-}(\mathcal{F}_d \times \mathcal{F}_d, \mathcal{E})$ into $L^2_{\text{fin}-}(\mathcal{F}_d \times \mathcal{F}_d, \mathcal{E}_*)$, and hence $L_T f$ is in the domain of W_* whenever $f \in L^2_{\text{fin}-}(\mathcal{F}_d \times \mathcal{F}_d, \mathcal{E})$. Hence L_T^{W,W_*} is well-defined as an operator from the dense subset $WL^2_{\text{fin}-}(\mathcal{F}_d \times \mathcal{F}_d, \mathcal{E})$ of $\mathcal{L}_W$ into $W_* L^2_{\text{fin}-}(\mathcal{F}_d \times \mathcal{F}_d, \mathcal{E}*) \subset \mathcal{L}_{W_*}$. Note that the space $L^2_{\text{arb}-}(\mathcal{F}_d \times \mathcal{F}_d, \mathcal{E})$ given by

$$L^2_{\text{arb}-}(\mathcal{F}_d \times \mathcal{F}_d, \mathcal{E}) := L(\mathcal{F}_d \times (\mathcal{F}_d \setminus \{\emptyset\}), \mathcal{E}_*) \oplus L^2(\mathcal{F}_d \times \{\emptyset\}, \mathcal{E}) \tag{2.4.9}$$

is the dual of $L^2_{\text{fin}-}(\mathcal{F}_d \times \mathcal{F}_d, \mathcal{E})$ with respect to the formal L^2-inner product. Using this duality, we see that the operator

$$L_T^{[*]} W_* L_T \colon L^2_{\text{fin}-}(\mathcal{F}_d \times \mathcal{F}_d, \mathcal{E}) \to L^2_{\text{arb}-}(\mathcal{F}_d \times \mathcal{F}_d, \mathcal{E})$$

is well defined, and hence $W_* L_T$, $L_T^{[*]}$ and $L_T^{[*]} W_* L_T$ are all well-defined Haplitz operators with domain including at least $L^2_{\text{fin}-}(\mathcal{F}_d \times \mathcal{F}_d, \mathcal{E})$ or $L^2_{\text{fin}-}(\mathcal{F}_d \times \mathcal{F}_d, \mathcal{E}_*)$ as appropriate. Formulas for the symbols of these respective Haplitz operators are

$$\widehat{W_* L_T}(z, \zeta) = \widehat{W}_*(z, \zeta) T(z), \tag{2.4.10}$$

$$\widehat{L_T^{[*]} W_*}(z, \zeta) = T(\zeta)^* \widehat{W}_*(z, \zeta), \tag{2.4.11}$$

$$\widehat{L_T^{[*]} W_* L_T}(z, \zeta) = T(\zeta)^* \widehat{W}_*(z, \zeta) T(z) + T(\zeta)^* k_{per}(z, \zeta) T(z). \tag{2.4.12}$$

These formulas are derived in [**BaV04**, Proposition 3.4] for the case where $T(z)$ is assumed to be a polynomial, and are actually special cases of more general formulas for the case where W and W_* are not necessarily Haplitz extensions of the identity; the case where $T(z)$ is assumed only to induce a bounded operator $M_T \colon L^2(\mathcal{F}_d, \mathcal{E}) \to L^2(\mathcal{F}_d, \mathcal{E}_*)$ follows easily from the polynomial case by an approximation argument.

Conversely, if W and W_* are positive semidefinite Haplitz extensions of the identity, then the map $f \mapsto Wf$ transforms $L^2(\mathcal{F}_d, \mathcal{E})$ unitarily onto $\mathcal{H}_W$, i.e., $\mathcal{H}_W = WL^2(\mathcal{F}_d, \mathcal{E})$ (with no closure required). Similarly, $\mathcal{H}_{W_*} = W_* L^2(\mathcal{F}_d, \mathcal{E}_*)$.

Hence, if $S \in \mathcal{S}(W, W_*)$, then we define an operator $M\colon L^2(\mathcal{F}_d, \mathcal{E}) \to L^2(\mathcal{F}_d, \mathcal{E}_*)$ by $SWf = W_*[Mf]$ for $f \in L^2(\mathcal{F}_d, \mathcal{E}_*)$. It follows from the first of the intertwining relations (2.4.2) that $MS_j^R = S_j^R M$ for $j = 1, \dots, d$, so necessarily $M = M_T$ for a power series $T(z) = \sum_{v \in \mathcal{F}_d} T_v z^v$ and $\|M_T\| = \|S|_{\mathcal{H}_W}\| \leq 1$, i.e., $T(z) \in \mathcal{S}_{nc,d}(\mathcal{E}, \mathcal{E}_*)$ (see (2.4.1)). The intertwining relations (2.4.2) then force S to have the form $S = L_T^{W,W_*}$. We have arrived at the following.

PROPOSITION 2.4.1. *For the case where W and W_* are positive semidefinite Haplitz extensions of the identity, for any $S \in \mathcal{S}(W, W_*)$ there is a multiplier $T(z) = \sum_v T_v z^v$ in $\mathcal{S}_{nc,d}(\mathcal{E}, \mathcal{E}_*)$ for which $S = L_T^{W,W_*}$.*

However, unlike the case $d = 1$ where $\|S\| = \|S|_{\mathcal{H}_W}\|$, it is not a priori obvious when a given multiplier $T(z) \in \mathcal{S}_{nc,d}(\mathcal{E}, \mathcal{E}_*)$ induces an $S \in \mathcal{S}(W, W_*)$ with $S = L_T^{W,W_*}$; in principle, the answer depends on the particular choices W and W_* of positive semidefinite Haplitz extensions of the identity. The following theorem lists various criteria for this to happen.

THEOREM 2.4.2. *Suppose that W and W_* are positive semidefinite Haplitz extensions of the identity and that the power series $T(z) = \sum_{v \in \mathcal{F}_d} T_v z^v \in \mathcal{S}_{nc,d}(\mathcal{E}, \mathcal{E}_*)$, so the operator $L_T^{W,W_*}\colon Wf \mapsto W_* L_T f$ is well defined from $L^2_{fin-}(\mathcal{F}_d \times \mathcal{F}_d, \mathcal{E})$ to $W_* L^2_{fin-}(\mathcal{F}_d \times \mathcal{F}_d, \mathcal{E}_*)$. Then the following are equivalent:*

(1) *L_T^{W,W_*} extends to a well-defined contraction operator from $\mathcal{L}_W$ into $\mathcal{L}_{W_*}$.*
(2) *The Haplitz operator $L := W - L_T^{[*]} W_* L_T$ is positive semidefinite.*
(3) *The block Haplitz operator $\begin{bmatrix} W_* & W_* L_T \\ L_T^{[*]} W_* & W \end{bmatrix}$ is positive semidefinite.*
(4) *The symbol of L*

$$\widehat{L}(z,\zeta) = \widehat{W}(z,\zeta) - T(\zeta)^* \widehat{W}_*(z,\zeta) T(z) - \sum_{u \neq \emptyset} (z^{-1})^{u^\top} T(\zeta)^* T(z) (\zeta^{-1})^u \tag{2.4.13}$$

and its defect $D_{\widehat{L}}$

$$\begin{aligned} D_{\widehat{L}}(z,\zeta) =& D_{\widehat{W}}(z,\zeta) - T(\zeta)^* D_{\widehat{W}_*}(z,\zeta) T(z) \\ =& [W(z,\zeta) - \sum_{k=1}^{d} z_k^{-1} \widehat{W}(z,\zeta) \zeta_k^{-1}] \\ & - T(\zeta)^* [\widehat{W}_*(z,\zeta) - \sum_{k=1}^{d} z_k^{-1} \widehat{W}_*(z,\zeta) \zeta_k^{-1}] T(z) \end{aligned} \tag{2.4.14}$$

are both positive symbols.
(5) *Both the power series $\widehat{L}_0$ given by*

$$\widehat{L}_0(z,\zeta) = \widehat{W}(z,\zeta) - T(\zeta)^* \widehat{W}_*(z,\zeta) T(z) \tag{2.4.15}$$

and $D_{\widehat{L}}(z,\zeta)$ as in (2.4.14) are positive symbols.
(6) *Both the power series*

$$X(z,\zeta) = (\widehat{W}(z,\zeta) - I) - T(\zeta)^* (\widehat{W}_*(z,\zeta) - I) T(z) \tag{2.4.16}$$

and

$$X'(z,\zeta) := D_X(z,\zeta) + I - T(\zeta)^* T(z)$$
$$(2.4.17) \qquad = X(z,\zeta) - \sum_{k=1}^{d} z_k^{-1} X(z,\zeta)\zeta_k^{-1} + I - T(\zeta)^* T(z)$$

are positive symbols.

In case W and W_* are Cuntz weights, then $D_{\widehat{L}}(z,\zeta) = 0$ and $X'(z,\zeta) = 0$. Hence, in this case, any of (1)–(3) is equivalent to

4′ *The symbol $\widehat{L}(z,\zeta)$ given by (2.4.13) is a positive symbol.*
5′ *The power series $\widehat{L}_0(z,\zeta)$ given by (2.4.15) is a positive symbol*
6′ *The power series $X(z,\zeta)$ given by (2.4.16) is a positive symbol.*

Proof. For $f \in L^2_{\text{fin}-}(\mathcal{F}_d \times \mathcal{F}_d, \mathcal{E})$, we compute

$$\begin{aligned}\langle (I - (L_T^{W,W_*})^* L_T^{W,W_*}) Wf, Wf\rangle_{\mathcal{L}_W} &= \langle Wf, Wf\rangle_{\mathcal{L}_W} - \langle L_T^{W,W_*} Wf, L_T^{W,W_*} Wf\rangle_{\mathcal{L}_{W_*}} \\ &= \langle Wf, f\rangle_{L^2} - \langle W_* L_T f, L_T f\rangle_{L^2} \\ &= \langle Wf, f\rangle_{L^2} - \langle L_T^{[*]} W_* L_T f, f\rangle_{L^2} \\ &= \langle (W - L_T^{[*]} W_* L_T) f, f\rangle_{L^2}\end{aligned}$$

and conclude the equivalence of (1) and (2) in the statement of the theorem. From the standard Schur-complement computation

$$\begin{bmatrix} W_* & W_* L_T \\ L_T^{[*]} W_* & W \end{bmatrix} = \begin{bmatrix} I & 0 \\ L_T^{[*]} & I \end{bmatrix} \begin{bmatrix} W_* & 0 \\ 0 & W - L_T^{[*]} W L_T \end{bmatrix} \begin{bmatrix} I & L_T \\ 0 & I \end{bmatrix}$$

we see the equivalence of (2) and (3).

By (2.4.12) we see that $\widehat{L}(z,\zeta)$ given by (2.4.13) is the symbol for the defect operator $L = W - L_T^{[*]} W_* L_T$. By a result from [**BaV04**] discussed above (see (2.2.31)–(2.2.32)), we know that (2) is equivalent to both $\widehat{L}(z,\zeta)$ and $D_{\widehat{L}}(z,\zeta)$ being positive symbols. For $D_{\widehat{L}}(z,\zeta)$, we compute

$$\begin{aligned} D_{\widehat{L}}(z,\zeta) &= \widehat{L}(z,\zeta) - \sum_{k=1}^{d} z_k^{-1} \widehat{L}(z,\zeta)\zeta_k^{-1} \\ &= \widehat{W}(z,\zeta) - T(\zeta)^* \widehat{W}_*(z,\zeta) T(z) - \sum_{u \neq \emptyset} (z^{-1})^{u^\top} T(\zeta)^* T(z) (\zeta^{-1})^u \\ &\quad - \sum_{k=1}^{d} z_k^{-1} \widehat{W}(z,\zeta)\zeta_k^{-1} + \sum_{k=1}^{d} z_k^{-1} T(\zeta)^* \widehat{W}_*(z,\zeta) T(z)\zeta_k^{-1} \\ &\quad + \sum_{u\colon |u| \geq 2} (z^{-1})^{u^\top} T(\zeta)^* T(z) (\zeta^{-1})^u \\ &= \widehat{W}(z,\zeta) - \sum_{k=1}^{d} z_k^{-1} \widehat{W}(z,\zeta)\zeta_k^{-1} \\ &\quad - T(\zeta)^* [\widehat{W}_*(z,\zeta) - \sum_{k=1}^{d} z_k^{-1} \widehat{W}_*(z,\zeta)\zeta_k^{-1}] T(z) + \sum_{k=1}^{d} z_k^{-1} T(\zeta)^* T(z) \zeta_k^{-1} \end{aligned}$$

$$- \sum_{u:\, u \neq \emptyset} (z^{-1})^{u^\top} T(\zeta)^* T(z) (\zeta^{-1})^u + \sum_{u:\, |u| \geq 2} (z^{-1})^{u^\top} T(\zeta)^* T(z) (\zeta^{-1})^u$$
$$= D_{\widehat{W}}(z,\zeta) - T(\zeta)^* D_{\widehat{W}_*}(z,\zeta) T(z).$$

Thus $D_{\widehat{L}}(z,\zeta)$ collapses to (2.4.14) and the equivalence of (2) and (4) follows.

By the same principle (2.2.31)–(2.2.32) for positivity of a [*]-Haplitz operator in terms of its symbol as was used above, we know that (3) is equivalent to both $\widehat{M}(z,\zeta)$ and $D_{\widehat{M}}(z,\zeta)$ being positive symbols, where $\widehat{M}$ is the symbol for the block matrix Haplitz operator $M = \begin{bmatrix} W_* & W_* L_T \\ L_T^{[*]} W_* & W \end{bmatrix}$. One then computes

$$\widehat{M}(z,\zeta) = \begin{bmatrix} \widehat{W}_*(z,\zeta) & \widehat{W}_*(z,\zeta)T(z) \\ T(\zeta)^* \widehat{W}_*(z,\zeta) & \widehat{W}(z,\zeta) \end{bmatrix}$$
$$D_{\widehat{M}}(z,\zeta) = \begin{bmatrix} D_{\widehat{W}_*}(z,\zeta) & D_{\widehat{W}_*}(z,\zeta)T(z) \\ T(\zeta)^* D_{\widehat{W}_*}(z,\zeta) & D_{\widehat{W}}(z,\zeta) \end{bmatrix}.$$

From the Schur-complement factorization

$$\widehat{M}(z,\zeta) = \begin{bmatrix} I & 0 \\ T(\zeta)^* & I \end{bmatrix} \begin{bmatrix} \widehat{W}_*(z,\zeta) & 0 \\ 0 & \widehat{W}(z,\zeta) - T(\zeta)^* \widehat{W}_*(z,\zeta) T(z) \end{bmatrix} \begin{bmatrix} I & T(z) \\ 0 & I \end{bmatrix}$$

we see that the symbol-positivity of $\widehat{M}(z,\zeta)$ is equivalent to the symbol positivity of $\widehat{W}(z,\zeta) - T(\zeta)^* \widehat{W}_*(z,\zeta) T(z) = \widehat{L}_0(z,\zeta)$, and similarly the symbol-positivity of $D_{\widehat{M}}(z,\zeta)$ is equivalent to the symbol-positivity of $D_{\widehat{W}}(z,\zeta) - T(\zeta)^* D_{\widehat{W}_*}(z,\zeta) T(z) = D_{\widehat{L}}(z,\zeta)$. Thus the equivalence of (3) and (5) follows.

As we have shown (2) $\iff$ (3), (2) $\iff$ (4) and (3) $\iff$ (5), the equivalence (4) $\iff$ (5) follows. It is instructive to see also how (4) $\iff$ (5) can be seen directly. Indeed, the symbol-positivity of $\widehat{L}$ trivially implies the symbol-positivity of $\widehat{L}_0$ (since the term $\sum_{u:\, u \neq \emptyset} (z^{-1})^{u^\top} T(\zeta)^* T(z) (\zeta^{-1})^u$ is positive), and hence (4) implies (5). Conversely, assume (5). Then, from the positivity of $D_{\widehat{L}}$ we see that

$$\begin{aligned}
&\widehat{W}(z,\zeta) - T(\zeta)^* \widehat{W}_*(z,\zeta) T(z) \\
&\geq \sum_{k=1}^d z_k^{-1} \widehat{W}(z,\zeta) \zeta_k^{-1} - T(\zeta)^* \left(\sum_{k=1}^d z_k^{-1} \widehat{W}_*(z,\zeta) \zeta_k^{-1} \right) T(z) \\
&= \sum_{k=1}^d z_k^{-1} \widehat{W}(z,\zeta) \zeta_k^{-1} - \sum_{k=1}^d z_k^{-1} \left(T(\zeta)^* \widehat{W}_*(z,\zeta) T(z) \right) \zeta_k^{-1} + \sum_{k=1}^d z_k^{-1} T(\zeta)^* T(z) \zeta_k^{-1} \\
&= \sum_{k=1}^d z_k^{-1} \left(\widehat{W}(z,\zeta) - T(\zeta)^* \widehat{W}_*(z,\zeta) T(z) \right) \zeta_k^{-1} + \sum_{k=1}^d z_k^{-1} T(\zeta)^* T(z) \zeta_k^{-1} \\
&\geq \sum_{u:\, |u|=2} (z^{-1})^{u^\top} \left(\widehat{W}(z,\zeta) - T(\zeta)^* \widehat{W}_*(z,\zeta) T(z) \right) (\zeta^{-1})^u \\
&\qquad + \sum_{u:\, 0<|u|\leq 2} (z^{-1})^{u^\top} T(\zeta)^* T(z) (\zeta^{-1})^u
\end{aligned}$$

and continuation of iteration of these inequalities leads to

$$\widehat{W}(z,\zeta) - T(\zeta)^*\widehat{W}_*(z,\zeta)T(z) \geq \sum_{u\colon 0<|u|\leq N} (z^{-1})^{u^\top} T(\zeta)^*T(z)(\zeta^{-1})^u$$

for all $N \in \mathbb{N}$, and it follows that $\widehat{L}(z,\zeta)$ is a positive symbol. In this way we see that (5) implies (4) and our direct proof of (4) $\iff$ (5) is complete.

We see that (5) implies (6) as follows. From the definitions for the expressions $\widehat{L}_0(z,\zeta)$ and $X(z,\zeta)$ (see (2.4.15) and (2.4.16)), we have the identity

$$X(z,\zeta) = \widehat{L}_0(z,\zeta) + T(\zeta)^*T(z) - I. \tag{2.4.18}$$

Let us isolate the terms of the form $z^v\zeta^\emptyset$ and $z^\emptyset\zeta^v$ in $\widehat{L}_0(z,\zeta)$ as follows:

$$\widehat{L}_0(z,\zeta) = I + T_\emptyset^*T_\emptyset - T_\emptyset^*T(z) - T(\zeta)^*T_\emptyset + \sum_{v,w\neq\emptyset} \Omega_{v,w}z^v\zeta^w. \tag{2.4.19}$$

From the fact that $\widehat{L}_0(z,\zeta)$ is positive, it follows that $\Omega(z,\zeta) := \sum_{v,w\neq\emptyset}\Omega_{v,w}z^v\zeta^w$ is positive as well. Now substitute (2.4.19) into (2.4.18) to write $X(z,\zeta)$ as

$$\begin{aligned}
X(z,\zeta) &= [I + T_\emptyset^*T_\emptyset - T_\emptyset^*T(z) - T(\zeta)^*T_\emptyset + \Omega(z,\zeta)] \\
&\qquad + [(T(\zeta) - T_\emptyset)^*(T(z) - T_\emptyset) - T_\emptyset^*T_\emptyset + T_\emptyset^*T(z) + T(\zeta)^*T_\emptyset - I] \\
&= \Omega(z,\zeta) + (T(\zeta) - T_\emptyset)^*(T(z) - T_\emptyset).
\end{aligned}$$

Thus $X(z,\zeta)$ is the sum of two positive symbols, and hence is a positive symbol. Furthermore, a computation gives

$$\begin{aligned}
D_X(z,\zeta) &= X(z,\zeta) - \sum_{k=1}^d z_k^{-1}X(z,\zeta)\zeta_k^{-1} \\
&= \left(\widehat{W}(z,\zeta) - I\right) - T(\zeta)^*\left(\widehat{W}_*(z,\zeta) - I\right)T(z) \\
&\qquad - \sum_{k-1}^d z_k^{-1}\widehat{W}(z,\zeta)\zeta_k^{-1} - \sum_{k=1}^d z_k^{-1}T(\zeta)^*\left(\widehat{W}_*(z,\zeta) - I\right)T(z)\zeta_k^{-1} \\
&= \widehat{W}(z,\zeta) - \sum_{k=1}^d z_k^{-1}\widehat{W}(z,\zeta)\zeta_k^{-1} - I \\
&\qquad - T(\zeta)^*\left(\widehat{W}_*(\zeta,z) - \sum_{k=1}^d z_k^{-1}\widehat{W}_*(z,\zeta)\zeta_k^{-1}\right)T(z) + T(\zeta)^*T(z) \\
&= D_{\widehat{L}}(z,\zeta) - I + T(\zeta)^*T(z).
\end{aligned} \tag{2.4.20}$$

Hence the symbol-positivity of $D_{\widehat{L}}(z,\zeta)$ implies the symbol-positivity of $X'(z,\zeta) = D_X(z,\zeta) + I - T(\zeta)^*T(z)$. This concludes the proof of (5) $\Longrightarrow$ (6).

Conversely, suppose that (6) holds, so both $X(z,\zeta)$ and $X'(z,\zeta)$ in (2.4.16) and (2.4.17) are positive symbols. From (2.4.20) we see that symbol-positivity of $X'(z,\zeta)$ leads to symbol-positivity of $D_{\widehat{L}}(z,\zeta)$. Furthermore, from (2.4.18) and the assumed positivity of $X'(z,\zeta)$ we have

$$\begin{aligned}\widehat{L}_0(z,\zeta) &= X(z,\zeta) + I - T(\zeta)^*T(z)\\ &\geq \sum_{k=1}^{d} z_k^{-1} X(z,\zeta)\zeta_k^{-1}\\ &\geq 0 \text{ (since } X(z,\zeta) \text{ is positive).}\end{aligned}\tag{2.4.21}$$

This completes the proof of (6) $\Longrightarrow$ (5).

Next, assume that W and W_* are Cuntz weights, so $D_{\widehat{W}}(z,\zeta) = 0$ as well as $D_{\widehat{W}_*}(z,\zeta) = 0$. An immediate consequence of the formula (2.4.14) is that then $D_{\widehat{L}}(z,\zeta) = 0$. From (2.4.20) we see that this then implies that $D_X(z,\zeta) = -I + T(\zeta)^*T(z)$, i.e., $X'(z,\zeta) = 0$. Thus, the conditions (4), (5), (6) in the statement of the theorem simplify to (4′), (5′) and (6′), as indicated, and the Theorem follows. □

As a corollary we obtain the following explicit formula for the action of $(L_T^{W,W_*})^*$ in the situation of Theorem 2.4.2.

COROLLARY 2.4.3. *Suppose that T, W, W_* are as in Theorem 2.4.2, so*

$$L_T^{W,W_*}\colon Wp \mapsto W_* L_T p \textit{ for } p \in L^2_{\mathit{fin}-}(\mathcal{F}_d \times \mathcal{F}_d, \mathcal{E})$$

extends to define a contraction operator from $\mathcal{L}_W$ into $\mathcal{L}_{W_*}$. Then $L_T^{[*]}$ maps $\mathcal{L}_{W_*}$ into $\mathcal{L}_W$ and the adjoint $(L_T^{W,W_*})^*\colon \mathcal{L}_{W_*} \to \mathcal{L}_W$ of L_T^{W,W_*} is given by

$$(L_T^{W,W_*})^* = L_T^{[*]}|_{\mathcal{L}_{W_*}}.$$

PROOF. By (2.2.23) (or the more abstract version (2.2.11)), we know that there is an auxiliary Hilbert space $\mathcal{K}$ and an operator $\Phi\colon \mathcal{K} \to L(\mathcal{F}_d \times \mathcal{F}_d, \mathcal{E})$ so that $W = \Phi\Phi^*$ and $\mathcal{L}_W = \Phi\mathcal{K}$. Similarly, there is an auxiliary Hilbert space $\mathcal{K}_*$ and an operator $\Phi_*\colon \mathcal{K}_* \to L(\mathcal{F}_d \times \mathcal{F}_d, \mathcal{E}_*)$ so that $\mathcal{L}_{W_*} = \Phi_*\mathcal{K}_*$. From condition (2) in Theorem 2.4.2 we read off that

$$\|\Phi_*^* L_T p\|^2_{\mathcal{K}_*} \leq \|\Phi^* p\|^2_{\mathcal{K}} \text{ for all } p \in \mathcal{P}(\mathcal{F}_d \times \mathcal{F}_d, \mathcal{E}).$$

Hence we may define a unique contraction operator $C\colon \mathcal{K} \to \mathcal{K}_*$ so that

$$C\colon \Phi^* p \mapsto \Phi_*^* L_T p \text{ for } p \in \mathcal{P}(\mathcal{F}_d \times \mathcal{F}_d, \mathcal{E}), \qquad C|_{\ker \Phi} = 0.$$

Taking adjoints on both sides of the operator equation $C\Phi^* = \Phi_*^* L_T$ then gives

$$L_T^{[*]}\Phi_* = \Phi C^* \text{ on } \mathcal{K}_*.$$

As $\mathcal{L}_{W_*} = \Phi_*\mathcal{K}_*$ and $\mathcal{L}_W = \Phi\mathcal{K}$, we conclude that $L_T^{[*]}$ maps $\mathcal{L}_{W_*}$ into $\mathcal{L}_W$ as asserted.

We next compute, for $p \in L^2_{\text{fin}-}(\mathcal{F}_d \times \mathcal{F}_d, \mathcal{E}_*)$ and $q \in L^2_{\text{fin}-}(\mathcal{F}_d \times \mathcal{F}_d, \mathcal{E})$,

$$\begin{aligned}\langle (L_T^{W,W_*})^* W_* p, Wq\rangle_{\mathcal{L}_W} &= \langle W_* p, L_T^{W,W_*} Wq\rangle_{\mathcal{L}_{W_*}} = \langle W_* p, W_* L_T q\rangle_{\mathcal{L}_{W_*}}\\ &= \langle W_* p, L_T q\rangle_{L^2}\\ &= \langle L_T^{[*]} W_* p, q\rangle_{L^2}\\ &= \langle L_T^{[*]} W_* p, Wq\rangle_{\mathcal{L}_W}\end{aligned}$$

where the last step is justified since we are guaranteed that $L_T^{[*]} W_* p \in \mathcal{L}_W$. From this computation we conclude that $(L_T^{W,W_*})^* = L_T^{[*]}|_{\mathcal{L}_{W_*}}$ as asserted □

REMARK 2.4.4. The formula (2.4.16) for $X(z,\zeta)$ has the following operator-theoretic interpretation. Given a Cuntz weight W, define a subspace $\mathcal{L}_{W,spp}$ (the "strict pure past" of $\mathcal{L}_W$) by

$$\mathcal{L}_{W,spp} = \text{closed span}\{\mathcal{U}_W^{*w} We \colon e \in \mathcal{E},\ w \in \mathcal{F}_d \setminus \{\emptyset\}\}.$$

It can be shown that the map $\pi\colon f(z,\zeta) \mapsto f(0,\zeta)$, or more precisely in terms of coefficients,

$$\pi\colon \sum_{v,w} f_{v,w} z^v \zeta^w \mapsto \sum_w f_{\emptyset,w}\zeta^w,$$

is injective on $\mathcal{L}_{W,spp}$. Define a space $\mathcal{H}(\widehat{W}-I)$ by

$$\mathcal{H}(\widehat{W}-I) = \{g(\zeta) = \sum_w g_w \zeta^w \colon g(\zeta) = f(0,\zeta) \text{ for some } f \in \mathcal{L}_{W,spp}\}$$

with norm

$$\|g\|_{\mathcal{H}(\widehat{W}-I)} = \|f\|_{\mathcal{H}_{W,spp}} \text{ if } g(\zeta) = f(0,\zeta) \text{ with } f \in \mathcal{H}_{W,spp}.$$

Then it can be shown that the condition $X(z,\zeta)$ being a positive symbol is simply an expression of the condition that multiplication by $T(\zeta)^* = \sum_{w\in\mathcal{F}_d} T^*_{w^\top}\zeta^w$ is a contraction operator from $\mathcal{H}(\widehat{W}_*-I)$ to $\mathcal{H}(\widehat{W}-I)$. Moreover, from Corollary 2.4.3 we see that $(L_T^{W,W_*})^* = L_T^{[*]}|_{\mathcal{L}_{W_*}}$ and, from the formula (2.4.7), we see that the pull-back of multiplication by $T(\zeta)^*$ on $\mathcal{H}(\widehat{W}_*-I)$ to $\mathcal{L}_{W_*,spp}$ is equal to $L_T^{[*]}|_{\mathcal{L}_{W_*,spp}}$, and hence to $(L_T^{W,W_*})^*|_{\mathcal{L}_{W,spp}}\colon \mathcal{L}_{W_*,spp} \to \mathcal{L}_{W,spp}$. Thus, the content of (1) $\Longleftrightarrow$ (6′) in Theorem 2.4.2 is that: $\|L_T^{W,W_*}\| \le 1 \Longleftrightarrow \|(L_T^{W,W_*})^*|_{\mathcal{L}_{W_*,spp}}\| \le 1$. The direction $\Longrightarrow$ is trivial. The converse direction $\Longleftarrow$ can also be seen directly using the fact that $\{\mathcal{U}^v_{W_*}\mathcal{L}_{W_*,spp} \colon v \in \mathcal{F}_d\}$ is dense in $\mathcal{L}_{W_*}$. In this way we get an independent, operator-theoretic proof of (1) $\Longleftrightarrow$ (6′) in Theorem 2.4.2.

Let us say that a triple (T,W,W_*) (where T is a noncommutative formal power series with coefficients in $\mathcal{L}(\mathcal{E},\mathcal{E}_*)$ and W and W_* are positive semidefinite Haplitz operators on $\mathcal{E}$ and $\mathcal{E}_*$ respectively) is *admissible* if $S := L_T^{W,W_*}$ is a contraction operator from $\mathcal{L}_W$ to $\mathcal{L}_{W_*}$. We seek to understand to what extent T,W determine W_* or T,W_* determine W for an admissible triple (T,W,W_*). The following result gives an answer.

THEOREM 2.4.5. *Let the noncommutative formal power series*

$$T(z) = \sum_{v\in\mathcal{F}_d} T_v z^v \ (\textit{with } T_v \in \mathcal{L}(\mathcal{E},\mathcal{E}_*))$$

and a positive semidefinite Haplitz extension W_* of the identity $[\delta_{w,\alpha} I_{\mathcal{E}_*}]$ be given.

(1) *Then positive semidefinite Haplitz operators W for which $L_T^{W,W_*}\colon \mathcal{L}_W \to \mathcal{L}_{W_*}$ is contractive (i.e., for which (T,W,W_*) is admissible) are in one-to-one correspondence with positive semidefinite Haplitz extensions L of the Cuntz-Toeplitz operator $I - M_T^* M_T$ on $L^2(\mathcal{F}_d \times \{\emptyset\},\mathcal{E})$ according to the formula*

$$W = L + L_T^{[*]} W_* L_T. \tag{2.4.22}$$

Moreover, if W_* is a Cuntz weight, then W is a Cuntz weight if and only if L is a Cuntz weight.

(2) *Positive semidefinite Haplitz operators W such that L_T^{W,W_*} is contractive (i.e., for which (T, W, W_*) is admissible) are in one-to-one correspondence with symbol-positive formal power series $X(z,\zeta)$ of the form*

$$X(z,\zeta) = \sum_{i,j=1}^{d} \zeta_j X^{ij}(z,\zeta) z_i \tag{2.4.23}$$

for which also

$$\begin{aligned} X'(z,\zeta) :=& D_X(z,\zeta) + I - T(\zeta)^* T(z) \\ :=& X(z,\zeta) - \sum_{k=1}^{d} z_k^{-1} X(z,\zeta)\zeta_k^{-1} + I - T(\zeta)^* T(z) \end{aligned} \tag{2.4.24}$$

is a positive symbol, according to the formula

$$\widehat{W}(z,\zeta) = X(z,\zeta) + I + T(\zeta)^*[\widehat{W}_*(z,\zeta) - I]T(z). \tag{2.4.25}$$

Moreover, given that W_* is a Cuntz weight, then W is a Cuntz weight if and only if $X'(z,\zeta) = 0$.

(3) *When (T, W, W_*) is an admissible triple with W and W_* equal to Cuntz weights, and L and X are given as in (2.4.22) and (2.4.25), then L and X uniquely determine each other via the following formula:*

$$\widehat{L}(z,\zeta) = X(z,\zeta) + I - T(\zeta)^* T(z) - \sum_{v' \neq \emptyset} (z^{-1})^{v'^{\top}} T(\zeta)^* T(z) (\zeta^{-1})^{v'}. \tag{2.4.26}$$

PROOF. Suppose that (T, W, W_*) is an admissible triple. If we define L by $L = W - L_T^{[*]} W_* L_T$, the fact that W and W_* are Haplitz extensions of the identity implies that L is a Haplitz extension of the Cuntz-Toeplitz operator $I - M_T^* M_T$. By (1) $\iff$ (2) in Theorem 2.4.2 we know that $L := W - L_T^{[*]} W_* L_T$ is positive-semidefinite; hence L is a positive-semidefinite Haplitz extension of $I - M_T^* M_T$. Conversely, if L is any positive semidefinite extension of $I - M_T^* M_T$, use (2.4.22) to define W:

$$W = L + L_T^{[*]} W_* L_T.$$

Clearly, W, as a sum of positive semidefinite Haplitz operators, is positive semidefinite. From $W - L_T^{[*]} W_* L_T = L \geq 0$, we see that L_T^{W,W_*} is contractive. From the formula (2.4.14) for $D_{\widehat{L}}(z,\zeta)$, given that W_* is a Cuntz weight (i.e., $D_{\widehat{W}_*}(z,\zeta) = 0$), we see that W is a Cuntz weight (i.e., $D_{\widehat{W}}(z,\zeta) = 0$) if and only if L is a Cuntz weight (i.e., $D_{\widehat{L}}(z,\zeta) = 0$) and assertion (1) follows.

Given that (T, W, W_*) is admissible, define $X(z,\zeta)$ via formula (2.4.25):

$$X(z,\zeta) = [\widehat{W}(z,\zeta) - I] - T(\zeta)^*[\widehat{W}_*(z,\zeta) - I]T(z).$$

By (1) $\iff$ (6) in Proposition 2.4.2, we see that both $X(z,\zeta)$ and $X'(z,\zeta)$ are positive symbols. By inspection we see that X has no terms of the form $z^{\emptyset}\zeta^w$ or $z^w\zeta^{\emptyset}$. From (2.4.20) we see that $X'(z,\zeta) = D_{\widehat{L}}(z,\zeta)$; hence $X'(z,\zeta) = 0$ exactly when $D_{\widehat{L}}(z,\zeta) = 0$; again from the formula (2.4.14), given that W_* is a Cuntz weight, this happens exactly when W is also a Cuntz weight.

Conversely, suppose that $X(z,\zeta)$ is a symbol-positive formal power series with no terms of the form $z^{\emptyset}\zeta^w$ or $z^w\zeta^{\emptyset}$ for which $X'(z,\zeta)$ is also symbol-positive. Use formula (2.4.25) to define W:

$$\widehat{W}(z,\zeta) = X(z,\zeta) + I + T(\zeta)^*[\widehat{W}_*(z,\zeta) - I]T(z). \tag{2.4.27}$$

Since W_* is a positive-definite Haplitz operator, we know that $D_{\widehat{W}_*}(z,\zeta) \geq 0$, or

$$\sum_{k=1}^{d} z_k \widehat{W}_*(z,\zeta)\zeta_k^{-1} \leq \widehat{W}_*(z,\zeta).$$

From the assumption that $X'(z,\zeta) \geq 0$, we see that

$$\sum_{k=1}^{d} z_k^{-1} X(z,\zeta)\zeta_k^{-1} \leq X(z,\zeta) + I - T(\zeta)^* T(z).$$

Hence we compute

$$\begin{aligned}
\sum_{k=1}^{d} z_k^{-1}\widehat{W}(z,\zeta)\zeta_k^{-1} &= \sum_{k=1}^{d} z_k^{-1} X(z,\zeta)\zeta_k^{-1} + T(\zeta)^* \left(\sum_{k=1}^{d} z_k^{-1}\widehat{W}_*(z,\zeta)\zeta_k^{-1}\right) T(z) \\
&\leq \sum_{k=1}^{d} z_k^{-1} X(z,\zeta)\zeta_k^{-1} + T(\zeta)^*\widehat{W}_*(z,\zeta)T(z) \\
&\leq X(z,\zeta) + I - T(\zeta)^*T(z) + T(\zeta)^*\widehat{W}_*(z,\zeta)T(z) \\
&= X(z,\zeta) + I + T(\zeta)^*[\widehat{W}_*(z,\zeta) - I]T(z) \\
&= \widehat{W}(z,\zeta) \qquad (2.4.28)
\end{aligned}$$

and we conclude that $D_{\widehat{W}}(z,\zeta) \geq 0$. Next use (2.4.27) to express X' in terms of (T, W, W_*) with X eliminated:

$$\begin{aligned}
X'(z,\zeta) &= D_X(z,\zeta) + I - T(\zeta)^*T(z) \\
&= \Bigg[\widehat{W}(z,\zeta) - I - T(\zeta)^*(\widehat{W}_*(z,\zeta) - I)T(z) \\
&\quad - \sum_{k=1}^{d} z_k^{-1}\widehat{W}(z,\zeta)\zeta_k^{-1} + T(\zeta)^*\left(\sum_{k=1}^{d} z_k^{-1}\widehat{W}_*(z,\zeta)\zeta_k^{-1}\right)T(z)\Bigg] \\
&\quad + I - T(\zeta)^*T(z) \\
&= \left[D_{\widehat{W}}(z,\zeta) - T(\zeta)^*D_{\widehat{W}_*}(z,\zeta)T(z) - I + T(\zeta)^*T(z)\right] + I - T(\zeta)^*T(z) \\
&= D_{\widehat{W}}(z,\zeta) - T(\zeta)^*D_{\widehat{W}_*}(z,\zeta)T(z) =: D_{\widehat{L}}(z,\zeta).
\end{aligned}$$

Hence positivity of $D_{\widehat{L}}(z,\zeta)$ follows from the assumed positivity of $X'(z,\zeta)$. Finally the same computation (2.4.21) used in the proof of (6) $\Longrightarrow$ (5) in Theorem 2.4.2 shows that positivity of $X(z,\zeta)$ and $X'(z,\zeta)$ leads to positivity of $\widehat{L}_0(z,\zeta)$. Hence statement (5) in Theorem 2.4.2 is verified, and it follows that (T, W, W_*) is admissible as desired. Next, given that W_* is a Cuntz weight, it is easily checked that equality holds in (2.4.28) if and only if $X'(z,\zeta) = 0$.

Finally, the formula (2.4.26) relating $\widehat{L}(x,\zeta)$ and $X(z,\zeta)$ falls out of the expressions (2.4.13) and (2.4.16). The completes the proof of Theorem 2.4.5. □

The multiplication operators M_T can be used to construct the Wold decomposition for a row isometry $S = (S_1, \ldots, S_d)$ and the analogue of Szegö factorization for this noncommutative setting. Let $W^+ = [W^+_{v,\alpha}]_{v,\alpha\in\mathcal{F}_d}$ be a positive semidefinite Toeplitz weight with associated row isometry $\mathcal{S}_{W^+} = (\mathcal{S}_{W^+,1} \ldots, \mathcal{S}_{W^+,d})$ as in (2.3.6).

The following result gives the Wold decomposition for the row isometry $S_{W^+} = (S_{W^+,1} \dots, S_{W^+,d})$ in terms of a "maximal factorable minorant" for the Toeplitz weight W^+.

THEOREM 2.4.6. *Let W^+ be a positive semidefinite Cuntz-Toeplitz weight on $\mathcal{E}$. Then there exists a Hilbert space $\mathcal{G}_1$ and a multiplication operator $M_{T_1}: \mathcal{P}(\mathcal{F}_d, \mathcal{E}) \to L^2(\mathcal{F}_d, \mathcal{G}_1)$ with $M_{T_1}\mathcal{P}(\mathcal{F}_d, \mathcal{E})$ dense in $L^2(\mathcal{F}_d, \mathcal{G}_1)$ such that*

$$W^+ \geq (M_{T_1})^* M_{T_1} \tag{2.4.29}$$

and, whenever $M_T: \mathcal{P}(\mathcal{F}_d, \mathcal{G}) \to L(\mathcal{F}_d, \mathcal{E})$ is another multiplication operator such that $W^+ \geq (M_T)^* M_T$, then we also have

$$(M_{T_1})^* M_{T_1} \geq (M_T)^* M_T. \tag{2.4.30}$$

These properties determine T_1 uniquely up to a unitary constant factor on the left.

Moreover, the following assertions are equivalent:

(1) *Equality holds in (2.4.29), i.e. $W^+ = (M_{T_1})^* M_{T_1}$.*
(2) *$\bigcap_{N\in\mathbb{N}} span_{v\in\mathcal{F}_d:\,|v|=N} \mathcal{S}^v_{W_+}\mathcal{H}^+_{W^+} = \{0\}$, i.e., $\mathcal{S}_{W^+}$ is a row shift.*
(3) *Polynomials are dense in $\mathcal{H}^+_{W^+}$.*

At the other extreme, we have the following set of equivalent conditions:

1′. *$T_1 = 0$.*
2′. *$\mathcal{S}_{W^+}$ is row unitary, i.e., for some (and hence for any) $N = 1, 2, 3, \dots,$*

$$\operatorname*{span}_{v:\,|v|=N} \mathcal{S}^v_{W^+}\mathcal{H}^+_{W_+} = \mathcal{H}^+_+.$$

3′. *$W^+\mathcal{P}_0(\mathcal{F}_d \setminus \{\emptyset\}, \mathcal{E})$ is dense in $\mathcal{H}^+_{W^+}$.*

PROOF. Let $\mathcal{H}^+_{W^+} = \mathcal{H}^+_{W^+,p} \oplus \mathcal{H}^+_{W^+,u}$ be the Wold decomposition for the row isometry $\mathcal{S}_{W^+} = (\mathcal{S}_{W^+,1}, \dots, \mathcal{S}_{W^+,d})$, i.e., $\mathcal{S}_{W^+}|_{\mathcal{H}^+_{W^+,p}}$ is a row shift, $\mathcal{S}_{W^+}|_{\mathcal{H}^+_{W^+,u}}$ is a row unitary (see [**Po89c**]). The pure part $\mathcal{H}^+_{W^+,p}$ is constructed as

$$\mathcal{H}^+_{W^+,p} = \oplus_{v\in\mathcal{F}_d}\mathcal{S}^v_{W^+}\mathcal{F}_1$$

where

$$\mathcal{F}_1 = \mathcal{H}^+_{W^+} \ominus \operatorname*{span}_{j=1,\dots,d} \mathcal{S}_{W^+,j}\mathcal{H}^+_{W^+}, \tag{2.4.31}$$

and then $\mathcal{H}^+_{W^+,u}$ can be defined as $\mathcal{H}^+_{W^+,u} = \mathcal{H}^+_{W^+} \ominus \mathcal{H}_{W^+,p}$ or equivalently as $\mathcal{H}_{W^+,u} = \bigcap_{N\in\mathbb{N}} \text{span}_{v\in\mathcal{F}_d:\,|v|=N} \mathcal{S}^v_{W^+}\mathcal{H}^+_{W^+}$. Set Φ_1 equal to the Fourier representation of the orthogonal projection from $\mathcal{H}^+_{W^+}$ onto $\mathcal{H}^+_{W^+,p}$, i.e., $\Phi_1: \mathcal{H}^+_{W^+} \to L^2(\mathcal{F}_d, \mathcal{G}_1)$ given by

$$\Phi_1(W^+p) = \sum_{v\in\mathcal{F}_d} \left(P_{\mathcal{G}_1}\mathcal{S}^{*v}_{W^+}W^+p\right) z^v.$$

Define a multiplier $T_1(z)$ by

$$T_1(z)e = \Phi_1 W^+[e](z) \text{ for } e \in \mathcal{E}.$$

Then

$$\begin{aligned}\Phi_1 W^+[z^v e] &= \Phi_1 W^+ S^{Rv^\top} e \\ &= \Phi_1 S_{W^+}^{v^\top} W^+ e \\ &= S^{Rv^\top} \Phi_1 W^+ e \\ &= S^{Rv^\top} M_{T_1} e \\ &= M_{T_1} S^{Rv^\top} e \\ &= M_{T_1}[z^v e]\end{aligned}$$

from which we see that $\Phi_1 W^+ = M_{T_1}$ on $\mathcal{P}(\mathcal{F}_d, \mathcal{E})$. In particular, it follows that $M_{T_1} \colon \mathcal{P}(\mathcal{F}_d, \mathcal{E}) \to L^2(\mathcal{F}_d, \mathcal{G}_1)$. Then, since

$$\begin{aligned}\|\Phi_1 W p\|^2_{L^2(\mathcal{F}_d, \mathcal{G}_1)} &= \|P_{\mathcal{H}^+_{W^+,p}} W p\|^2_{\mathcal{H}^+_{W^+}} \\ &\le \|W^+ p\|^2_{\mathcal{H}^+_{W^+}},\end{aligned}$$

we deduce that, for $p \in \mathcal{P}(\mathcal{F}_d, \mathcal{E})$,

$$\begin{aligned}\langle (W^+ - M^*_{T_1} M_{T_1}) p, p \rangle_{L^2} &= \langle W^+ p, p \rangle_{L^2} - \langle M_{T_1} p, M_{T_1} p \rangle_{L^2} \\ &= \|W^+ p\|^2_{\mathcal{H}^+_{W^+}} - \|\Phi_1 W^+ p\|^2_{L^2} \\ &= \|W^+ p\|^2_{\mathcal{H}^+_{W^+}} - \|P_{\mathcal{H}^+_{W^+,p}} W^+ p\|^2_{\mathcal{H}^+_{W^+}} \\ &= \|P_{\mathcal{H}^+_{W^+,u}} W^+ p\|^2 \ge 0\end{aligned}$$

and hence (2.4.29) holds as asserted. Moreover, equality holds in (2.4.29) if and only if $\mathcal{H}^+_{W^+,u} = \{0\}$, i.e., if and only if

$$\bigcap_{N \in \mathbb{N}} \operatorname{span}_{v \in \mathcal{F}_d \colon |v| = N} \mathcal{S}^v_{W^+} \mathcal{H}^+_{W^+} = \{0\}$$

and the equivalence of (1) and (2) follows.

We next show that (1) $\implies$ (3). Suppose that $W^+ = M^*_{T_1} M_{T_1}$ with T_1 outer (i.e., $M_{T_1} \mathcal{P}(\mathcal{F}_d, \mathcal{E})$ is dense in $L^2(\mathcal{F}_d, \mathcal{G}_1)$). Then from

$$\begin{aligned}\langle (M_{T_1})^* M_{T_1} f, (M_{T_1})^* M_{T_1} f \rangle_{\mathcal{H}^+_{W^+}} &= \langle W^+ f, W^+ f \rangle_{\mathcal{H}^+_{W^+}} \\ &= \langle W^+ f, f \rangle_{L^2} \\ &= \langle M_{T_1} f, M_{T_1} f \rangle_{L^2} \end{aligned} \tag{2.4.32}$$

we see that $(M_{T_1})^*$ is an isometry from the dense subspace $M_{T_1} \mathcal{P}(\mathcal{F}_d, \mathcal{E})$ of $L^2(\mathcal{F}_d, \mathcal{E})$ to a dense subspace $(M_{T_1})^* M_{T_1} \mathcal{P}(\mathcal{F}_d, \mathcal{E}) = W^+ \mathcal{P}(\mathcal{F}_d, \mathcal{E})$ of $\mathcal{H}_{W^+}$. Hence $(M_{T_1})^*$ defines a unitary transformation of $L^2(\mathcal{F}_d, \mathcal{E})$ onto $\mathcal{H}_{W^+}$. It follows that the subspace $(M_{T_1})^* \mathcal{P}(\mathcal{F}_d, \mathcal{G}_1)$ is dense in $\mathcal{H}_{W^+}$. But since $M_{T_1} : \mathcal{P}(\mathcal{F}_d, \mathcal{G}_1) z^\alpha \to \mathcal{P}(\mathcal{F}_d, \mathcal{G}_1) z^\alpha$ for each $\alpha \in \mathcal{F}_d$, we see that $(M_{T_1})^*$ takes polynomials to polynomials, and hence the dense set $(M_{T_1})^* \mathcal{P}(\mathcal{F}_d, \mathcal{E})$ in $\mathcal{H}_{W^+}$ consists entirely of polynomials. In this way we see that (1) $\implies$ (3) holds.

We next show that (3) $\implies$ (2). We first note that if we factor W^+ as $W^+ = \Psi \Psi^{[*]}$ for some $\Psi \colon \mathcal{H} \to L(\mathcal{F}_d, \mathcal{E})$ (where $\Psi^{[*]}$ is the adjoint with respect to the $(L(\mathcal{F}_d, \mathcal{G}_1) - \mathcal{P}(\mathcal{F}_d, \mathcal{G}_1))$-pairing in the L^2 inner product on the range of Ψ and the Hilbert space $\mathcal{H}$-inner product on the domain of Ψ), then a computation similar to (2.4.32) shows that $\mathcal{H}^+_{W^+} = \Psi \mathcal{H}$. Since W^+ is Cuntz-Toeplitz, we have

$\sum_{j=1}^{d} S_j^{R[*]} W^+ S_j^R = W^+$ on $\mathcal{P}(\mathcal{F}_d, \mathcal{E})$. This leads to the alternative factorization $W^+ = \Psi' \Psi'^*$ where

$$\Psi' = \begin{bmatrix} S_1^{R[*]}\Psi & \dots & S_d^{R[*]}\Psi \end{bmatrix} : (\oplus_{j=1}^{d} \mathcal{H}) \to L(\mathcal{F}_d, \mathcal{E}).$$

Thus we also have $\mathcal{H}_{W^+}^+ = \Psi'(\oplus_{j=1}^{d}\mathcal{H})$. In particular, it follows that $S_j^{R[*]}\Psi'\mathcal{H} \subset \mathcal{H}_{W^+}^+$, or $S_j^{R[*]}$ maps $\mathcal{H}_{W^+}^+$ into itself, for each $j = 1, \dots, d$. Then the formal computation

$$\begin{aligned} \langle \mathcal{S}_{W^+,j} W^+ p, W^+ q \rangle_{\mathcal{H}_{W^+}^+} &= \langle W^+ S_j^R p, q \rangle_{L^2} \\ &= \langle p, S_j^{R[*]} W^+ q \rangle_{L^2} \\ &= \langle W^+ p, S_j^{R[*]} W^+ q \rangle_{\mathcal{H}_{W^+}^+} \end{aligned}$$

shows that

$$\mathcal{S}_{W^+,j}^* = S_j^{R[*]}|_{\mathcal{H}_{W^+}}. \tag{2.4.33}$$

Now suppose that the polynomials in $\mathcal{H}_{W^+}$ are dense in $\mathcal{H}_{W^+}$. Note that the polynomials can be identified with $\bigcup_{N=0}^{\infty} \bigcap_{v \in \mathcal{F}_d \colon |v|=N} \ker S^{R*v}$. Hence the density of polynomials in $\mathcal{H}_{W^+}^+$ combined with the fact (2.4.33) implies that

$$\bigcup_{N \in \mathbb{N}} \bigcap_{v \in \mathcal{F}_d \colon |v|=N} \ker \mathcal{S}_{W^+}^{*v} \text{ is dense in } \mathcal{H}_{W^+}^+.$$

But this latter set is exactly the orthogonal complement of the set appearing on the left hand side of condition (2). Since density of the polynomials implies that the orthogonal complement of the polynomials is trivial, we see that (2) follows.

Similarly, from the construction we see that $T_1 = 0$ if and only if $\mathcal{F}_1$ given by (2.4.31) is zero. It is straightforward next to deduce the equivalence of this condition with (2′) and (3′), and the equivalence of (1′)–(3′) follows.

Next suppose that $M_T \colon \mathcal{P}(\mathcal{F}_d, \mathcal{E}) \to L(\mathcal{F}_d, \mathcal{G})$ is another multiplication operator for which $M_T^* M_T \le W^+$. In particular, we see that, for $p \in \mathcal{P}(\mathcal{F}_d, \mathcal{E})$,

$$\|M_T p\|_{L^2}^2 \le \langle W^+ p, p \rangle_{L^2} < \infty$$

so we see that $M_T \colon \mathcal{P}(\mathcal{F}_d, \mathcal{E}) \to L^2(\mathcal{F}_d, \mathcal{G})$. Define an operator $\Phi \colon WL^2(\mathcal{F}_d, \mathcal{E}) \to L^2(\mathcal{F}_d, \mathcal{G})$ by

$$\Phi W p = M_T p \text{ for } p \in L^2(\mathcal{F}_d, \mathcal{E}).$$

Then, from the assumed inequality $M_T^* M_T \le W^+$, we see that Φ extends uniquely to a contraction operator $\Phi \colon \mathcal{H}_{W^+}^+ \to L^2(\mathcal{F}_d, \mathcal{G})$. Moreover, from

$$\begin{aligned} \Phi \mathcal{S}_{W^+,j} W^+ p &= \Phi W^+ S_j^R p \\ &= M_T S_j^R p \\ &= S_j^R M_T p \\ &= S_j^R \Phi W^+ p, \end{aligned}$$

we see that $\Phi \mathcal{S}_{W^+,j} = S_j^R \Phi$. Hence

$$\Phi \cdot \bigcap_{N \in \mathbb{N}} \operatorname{span}_{v \in \mathcal{F}_d \colon |v|=N} \mathcal{S}_{W^+}^v \mathcal{H}_{W^+}^+ \subset \bigcap_{N \in \mathbb{N}} \operatorname{span}_{v \in \mathcal{F}_d \colon |v|=N} S^{Rv} L^2(\mathcal{F}_d, \mathcal{G}) = \{0\}$$

and hence $\Phi P_{\mathcal{H}_{W^+,u}} = 0$ or $\Phi W^+ p = \Phi(P_{\mathcal{H}_{W^+,p}} W^+ p)$. We conclude that

$$\begin{aligned}\|M_T p\|_{L^2} &= \|\Phi(W^+ p)\|_{L^2} = \|\Phi(P_{\mathcal{H}^+_{W^+,p}} W^+ p\|_{L^2} \\ &\leq \|P_{\mathcal{H}^+_{W^+}} W^+ p\|_{\mathcal{H}^+_{W^+}} = \|\Phi_1 W^+ p\|_{\mathcal{H}^+_{W^+}} = \|M_{T_1} p\|_{L^2}\end{aligned}$$

from which we see that

$$M_T^* M_T \leq M_{T_1}^* M_{T_1}$$

and (2.4.30) follows.

It remains to show uniqueness. If $M_{T_1'} : \mathcal{P}(\mathcal{F}_d, \mathcal{E}) \to L(\mathcal{F}_d, \mathcal{G}_1')$ were another multiplication operator with $M_{T_1'}\mathcal{P}(\mathcal{F}_d, \mathcal{E})$ dense in $L^2(\mathcal{F}_d, \mathcal{G}_1')$ for which (2.4.29) and(2.4.30) hold, then necessarily

$$M_{T_1'}^* M_{T_1'} = M_{T_1}^* M_{T_1}.$$

We may then define a unitary $U\colon L^2(\mathcal{F}_d, \mathcal{G}_1) \to L^2(\mathcal{F}_d, \mathcal{G}_1')$ densely with dense image by

$$U\colon T_1 p \mapsto T_1' p.$$

As U intertwines S_j^R on $L^2(\mathcal{F}_d, \mathcal{G}_1)$ with S_j^R on $L^2(\mathcal{F}_d, \mathcal{G}_1')$, it follows that U preserves wandering subspaces, i.e., U mapst the space

$$\mathcal{G}_1 \cong L^2(\mathcal{F}_d, \mathcal{G}_1) \ominus \operatorname*{span}_{j=1,\dots,d} S_j^R L^2(\mathcal{F}_d, \mathcal{G}_1)$$

onto

$$\mathcal{G}_1' \cong L^2(\mathcal{F}_d, \mathcal{G}_1') \ominus \operatorname*{span}_{j=1,\dots,d} S_j^R L^2(\mathcal{F}_d, \mathcal{G}_1').$$

If we set $U_0 = U|_{\mathcal{G}_1} : \mathcal{G}_1 \to \mathcal{G}_1'$, then U_0 is a unitary constant and $M_{T_1'} = M_{U_0} M_{T_1}$ as asserted. This completes the proof of Theorem 2.4.6. □

This result can be used to reduce the problem of Cuntz-weight extension of a given positive-semidefinite Cuntz-Toeplitz weight W^+ to the special case where $W^+_{v,\alpha} = \delta_{v,\alpha} I_{\mathcal{E}}$.

THEOREM 2.4.7. *Let W^+ be a positive semidefinite Cuntz-Toeplitz weight with Wold decomposition*

$$W^+ = M_{T_1}^* M_{T_1} + L^+$$

where the multiplication operator M_{T_1} acts from $\mathcal{P}(\mathcal{F}_d, \mathcal{E})$ to $L(\mathcal{F}_d, \mathcal{G}_1)$ as in Theorem 2.4.6. Then Cuntz-weight extensions W of W^+ are in one-to-one correspondence with Cuntz-weight extensions V of $V^+ = I$ on $L^2(\mathcal{F}_d, \mathcal{G}_1)$ according to the formula

$$W = L_{T_1}^{[*]} V L_{T_1} + L$$

where L is the unique Cuntz extension of L^+. Hence, W^+ has a unique Cuntz-weight extension W if and only if $W^+ = L^+$, i.e., the maximal factorable minorant for W^+ is 0.

PROOF. Suppose that $W = L_{T_1}^{[*]} V L_{T_1} + L$ where L is a Cuntz-weight extension of I. From the extension property (2.4.3) of L_{T_1} it is clear that W extends $W^+ = M_{T_1}^* M_{T_1} + L^+$. From the intertwining properties (2.4.4)-(2.4.5) of L_{T_1} and the fact that V and L are $[*]$-Haplitz, it is clear that W is $[*]$-Haplitz. As V and L are positive-semidefinite, we see that W is positive-semidefinite. From Theorem 2.4.2 we see that $L_{T_1}^{W,V}\colon \mathcal{L}_W \to \mathcal{L}_V$ and $L_I^{W,L}\colon \mathcal{L}_W \to \mathcal{L}_L$ given by

$$L_{T_1}^{W,V}\colon Wp \mapsto VL_{T_1}p, \qquad L_I^{W,L}\colon Wp \mapsto Lp$$

are well-defined contraction operators which intertwine the respective row isometries and their adjoints:

$$\begin{aligned} L_{T_1}^{W,V}\mathcal{U}_{W,j} &= \mathcal{U}_{V,j}L_{T_1}^{W,V}, & L_{T_1}^{W,V}\mathcal{U}_{W,j}^* &= \mathcal{U}_{V,j}^*L_{T_1}^{W,V}, \\ L_I^{W,L}\mathcal{U}_{W,j} &= \mathcal{U}_{L,j}L_I^{W,L}, & L_I^{W,L}\mathcal{U}_{W,j}^* &= \mathcal{U}_{L,j}^*L_I^{W,L} \end{aligned}$$

for $j = 1, \ldots, d$. By the assumption that V and L are Cuntz weights, we know by the discussion in Section 2.2 that $\mathcal{U}_V = (\mathcal{U}_{V,1}, \ldots, \mathcal{U}_{V,d})$ and $\mathcal{U}_L = (\mathcal{U}_{L,1}, \ldots, \mathcal{U}_{L,d})$ are row-unitary. Moreover, the defining expression $W = L_{T_1}^{[*]}VL_{T_1} + L$ for W can be rewritten as

$$I_{\mathcal{L}_W} = (L_{T_1}^{W,V})^*(L_{T_1}^{W,V}) + (L_I^{W,L})^*(L_I^{W,L}).$$

Hence we have

$$\begin{aligned} \sum_{j=1}^d \mathcal{U}_{W,j}\mathcal{U}_{W,j}^* &= \sum_{j=1}^d \left(\mathcal{U}_{W,j}\left[(L_{T_1}^{W,V})^*(L_{T_1}^{W,V}) + (L_I^{W,L})^*(L_I^{W,L})\right]\mathcal{U}_{W,j}^*\right) \\ &= (L_{T_1}^{W,V})^*\left(\sum_{j=1}^d \mathcal{U}_{V,j}\mathcal{U}_{V,j}^*\right)(L_{T_1}^{W,V}) + (L_I^{W,L})^*\left(\sum_{j=1}^d \mathcal{U}_{L,j}\mathcal{U}_{L,j}^*\right)(L_I^{W,L}) \\ &= (L_{T_1}^{W,V})^*(L_{T_1}^{W,V}) + (L_I^{W,L})^*(L_I^{W,L}) \\ &= I_{\mathcal{L}_W}. \end{aligned}$$

Hence $\mathcal{U}_W$ is row-unitary. The equivalence between the row-unitary property of $\mathcal{U}_W$ and W being a Cuntz weight delineated in Section 2.2 now guarantees us that W is a Cuntz weight as asserted.

Suppose conversely that W is a Cuntz-weight extension of $W^+ = M_{T_1}^*M_{T_1} + L^+$. As explained in Proposition 2.3.4, $\mathcal{U}_W = (\mathcal{U}_{W,1}, \ldots, \mathcal{U}_{W,d})$ is a minimal row-unitary extension of $\mathcal{S}_{W^+} = (\mathcal{S}_{W^+,1}, \ldots, \mathcal{S}_{W^+,d})$ on $\mathcal{H}_{W^+}$ (where we use the map $W^+p \mapsto Wp$ to isometrically identify $\mathcal{H}_{W^+}^+$ as a subspace of $\mathcal{L}_W$). From the Wold decomposition $\mathcal{S}_{W^+} = \mathcal{S}_{M_{T_1}^*M_{T_1}} \oplus L_{L^+}$ for $\mathcal{S}_{W^+}$, we see that necessarily $\mathcal{U}_W$ has the form

$$\mathcal{U}_W = \mathcal{U}_{W^{(s)}} \oplus \mathcal{U}_L \tag{2.4.34}$$

where $W^{(s)}$ is a Haplitz extension of $M_{T_1}^*M_{T_1}$. By the model theory for row-unitaries from [**BaV04**] summarized in Section 2.2, we know that, after proper identifications, W has the form

$$W_{v,w;\alpha,\beta} = P_{\mathcal{E}}\mathcal{U}_W^w\mathcal{U}_W^{*v}\mathcal{U}_W^{\alpha^\top}\mathcal{U}_W^{*\beta^\top}|_{\mathcal{E}}.$$

Since $\mathcal{U}_W$ splits as in (2.4.34), we see that

$$W_{v,w;\alpha,\beta} = W_{v,w;\alpha,\beta}^{(s)} + L_{v,w;\alpha,\beta}.$$

Next, use the properties (2.4.3)–(2.4.5) of L_{T_1} and the fact that L_{T_1} has dense range to deduce that any positive-semidefinite Haplitz extension of $M_{T_1}^*M_{T_1}$ has the form $L_{T_1}^{[*]}VL_{T_1}$ for a uniquely determined positive-semidefinite Haplitz extension V of I.

To see that V must be a Cuntz weight if $W^{(s)}$ is Cuntz, note that $W^{(s)} = L_{T_1}^{[*]} V L_{T_1}$ can be rewritten as

$$I_{\mathcal{L}_{W^{(s)}}} = (L_{T_1}^{W^{(s)},V})^* (L_{T_1}^{W^{(s)},V})$$

where $L_{T_1} : \mathcal{L}_W \to \mathcal{L}_V$ is the associated intertwining map

$$L_{T_1}^{W^{(s)},V} : W^{(s)} p \mapsto V L_{T_1} p.$$

As $L_{T_1}^{W^{(s)},V}$ also has dense range in $\mathcal{L}_V$ (since by construction $M_{T_1} \mathcal{P}(\mathcal{F}_d, \mathcal{E})$ is dense in $L^2(\mathcal{F}_d, \mathcal{G}_1)$), we see that in fact $L_{T_1}^{W^{(s)},V} : \mathcal{L}_{W^{(s)}} \to \mathcal{L}_V$ is unitary. Now use the intertwining properties of $L_{T_1}^{W^{(s)},V}$

$$L_{T_1}^{W^{(s)},V} \mathcal{U}_{W,j} = \mathcal{U}_{V,j} L_{T_1}^{W^{(s)},V}, \qquad L_{T_1}^{W^{(s)},V} \mathcal{U}_{W^{(s)},j}^* = \mathcal{U}_{W^{(s)},j}^* L_{T_1}^{W^{(s)},V}$$

to deduce that

$$\begin{aligned} I_{\mathcal{L}_{W^{(s)}}} &= \sum_{j=1}^{d} \mathcal{U}_{W^{(s)},j} \mathcal{U}_{W^{(s)},j}^* \\ &= \sum_{j=1}^{d} \mathcal{U}_{W^{(s)},j} (L_{T_1}^{W^{(s)},V})^* (L_{T_1}^{W^{(s)},V}) \mathcal{U}_{W^{(s)}j}^* \\ &= (L_{T_1}^{W^{(s)},V})^* \left(\sum_{j=1}^{d} \mathcal{U}_{V,j} \mathcal{U}_{V,j}^* \right) (L_{T_1}^{W^{(s)},V}). \end{aligned}$$

Since $L_{T_1}^{W^{(s)},V}$ is unitary, we now see that $\mathcal{U}_V = (\mathcal{U}_{V,1}, \ldots, \mathcal{U}_{V,d})$ is row-unitary, i.e., that V is a Cuntz weight, and the Theorem follows. □

We have seen that the classification of all Cuntz-weight extensions of the identity in general is difficult; indeed, Proposition 2.3.5 is only a partial solution of this issue. Nevertheless, the free-atomic representations of the Cuntz algebra presented in [**DaP99**] (see also [**BaV04**]) give a whole explicit class of row-unitary extensions of a row shift, and therefore Cuntz-weight extensions of the identity. Theorem 2.4.7 put in conjunction with this fact then guarantees that *Cuntz-weight extensions W of a positive semidefinite Cuntz-Toeplitz operator W^+ always exist.* This remark leads us to the following.

COROLLARY 2.4.8. *Let $T(z) = \sum_{v \in \mathcal{F}_d} T_v z^v$ be an element of the noncommutative Schur-class $\mathcal{S}_{nc,d}(\mathcal{E}, \mathcal{E}_*)$ and let W_* be any Cuntz-weight extension of $W_*^+ = [\delta_{v,\alpha} I_{\mathcal{E}_*}]$. Then Cuntz-weight extensions W of $[\delta_{v,\alpha} I_{\mathcal{E}}]_{v,\alpha \in \mathcal{F}_d}$ such that $S = L_T^{W,W_*}$ is in the Laurent noncommutative Schur-class $\mathcal{S}(W, W_*)$ are in one-to-one correspondence with Cuntz-weight extensions L of $L^+ := I - M_T^* M_T$ according to the formula*

$$W = L + L_T^{[*]} W_* L_T.$$

In particular, given any $T(z) \in \mathcal{S}_{nc,d}(\mathcal{E}, \mathcal{E}_*)$, there always exists Cuntz weight extensions of the identity W and W_* such that L_T^{W,W_*} is in the Laurent Schur multiplier class $\mathcal{S}(W, W_*)$.

PROOF. Let W_* be any Cuntz-weight extension of $[W_{*;\alpha,\beta}^+] = [\delta_{v,\alpha} I_{\mathcal{E}_*}]$ and let L be any Cuntz-weight extension of $L^+ := I - M_T^* M_T$; note that existence of such extensions follows from the remark immediately preceding the statement of the corollary. Set $W = L + L_T^{[*]} W_* L_T$. Then we see from Theorem 2.4.5 (1) that $S = L_T^{W,W_*} : \mathcal{L}_W \to \mathcal{L}_{W_*}$ is contractive as desired. $\square$

CHAPTER 3

Cuntz Scattering Systems

3.1. Basic geometric structure of Cuntz scattering systems

We define a *Cuntz scattering system* to be a collection

$$\mathfrak{S} = (\mathcal{U} = (\mathcal{U}_1, \dots, \mathcal{U}_d); \mathcal{K}, \mathcal{G}, \mathcal{G}_*) \tag{3.1.1}$$

where $\mathcal{U} = (\mathcal{U}_1, \dots, \mathcal{U}_d)$ is a row-unitary d-tuple of operators on the Hilbert space $\mathcal{K}$ (the *ambient space*) and where $\mathcal{G}$ (the *outgoing space*) and $\mathcal{G}_*$ (the *incoming space*) are subspaces of $\mathcal{K}$ such that

(1) The outgoing space $\mathcal{G}$ is invariant for each of the operators $\mathcal{U}_1, \dots, \mathcal{U}_d$, and the restricted d-tuple $\mathcal{U}|_\mathcal{G} = (\mathcal{U}_1|_\mathcal{G}, \dots, \mathcal{U}_d|_\mathcal{G})$ is a row shift (i.e., a row isometry with the property (2.3.7)) on $\mathcal{G}$.

(2) The incoming space $\mathcal{G}_*$ is invariant for each of $\mathcal{U}_1^* \dots, \mathcal{U}_d^*$ and $\mathcal{G}_* \subset \widetilde{\mathcal{G}}_*$, where we have set $\widetilde{\mathcal{G}}_*$ equal to the smallest subspace of $\mathcal{K}$ reducing for each of $\mathcal{U}_1, \dots, \mathcal{U}_d$ and containing the subspace $\mathcal{E}_*$ (the *incoming wandering subspace*) given by

$$\mathcal{E}_* = [\oplus_{j=1}^d \mathcal{U}_j \mathcal{G}_*] \ominus \mathcal{G}_*.$$

(3) The outgoing space $\mathcal{G}$ is orthogonal to the incoming space $\mathcal{G}_*$:

$$\mathcal{G} \perp \mathcal{G}_*. \tag{3.1.2}$$

Given a Cuntz scattering system $\mathfrak{S}$, define the space $\mathcal{E}$ (the *outgoing wandering subspace*) by

$$\mathcal{E} = \mathcal{G} \ominus [\oplus_{j=1}^d \mathcal{U}_j \mathcal{G}].$$

As $\mathcal{U}|_\mathcal{G}$ is a row shift, $\mathcal{E}$ is wandering for $\mathcal{U}$ (i.e., $\mathcal{U}^w \mathcal{E} \perp \mathcal{U}^v \mathcal{E}$ for $u, v \in \mathcal{F}_d$ with $u \neq v$) and $\mathcal{G}$ has the orthogonal decomposition

$$\mathcal{G} = \oplus_{v \in \mathcal{F}_d} \mathcal{U}^v \mathcal{E}.$$

We denote the smallest closed subspace of $\mathcal{K}$ reducing for each of $\mathcal{U}_1, \dots, \mathcal{U}_d$ and containing the space $\mathcal{G}$ by $\widetilde{\mathcal{G}}$. Since $\mathcal{U}$ is row unitary, the space $\widetilde{\mathcal{G}}$ can be expressed more efficiently as

$$\widetilde{\mathcal{G}} = \text{closed span}\{\mathcal{U}^{v^\top} \mathcal{U}^{*w^\top} e \colon v, w \in \mathcal{F}_d \text{ and } e \in \mathcal{E}.\} \tag{3.1.3}$$

The analogous spaces $\widetilde{\mathcal{G}}_*$ and $\mathcal{E}_*$ associated with incoming data were already introduced in the definition of a Cuntz scattering system. The geometric structure of these spaces is summarized in the following proposition.

PROPOSITION 3.1.1. *Let $\mathfrak{S}$ be a Cuntz scattering system* (3.1.1) *with incoming space $\mathcal{G}_*$. Set $\mathcal{E}_* = [\oplus_{j=1}^d \mathcal{U}_j \mathcal{G}_*] \ominus \mathcal{G}_*$, and let $\widetilde{\mathcal{G}}_*$ be the smallest reducing subspace* *for each of* $\mathcal{U}_1, \dots, \mathcal{U}_d$ *containing* $\mathcal{E}_*$ *as above. Then* $\widetilde{\mathcal{G}}_*$ *can also be described as the smallest subspace reducing for each of* $\mathcal{U}_1, \dots, \mathcal{U}_d$ *containing* $\mathcal{G}_*$*:*

$$\begin{aligned}\widetilde{\mathcal{G}}_* &= \text{closed span}\{\mathcal{U}^{v^\top}\mathcal{U}^{*w^\top}e_* \colon v, w \in \mathcal{F}_d \text{ and } e_* \in \mathcal{E}_*\} \\ &= \text{closed span}\{\mathcal{U}^{v^\top}\mathcal{U}^{*w^\top}g_* \colon v, w \in \mathcal{F}_d \text{ and } g_* \in \mathcal{G}_*\}\end{aligned}$$

Moreover, $\mathcal{E}_*$ *is wandering for* $\mathcal{U}$ *and* $\widetilde{\mathcal{G}}_*$ *also can be described via the orthogonal decomposition*

$$\widetilde{\mathcal{G}}_* = \mathcal{G}_* \oplus [\oplus_{v \in \mathcal{F}_d}\mathcal{U}^w\mathcal{E}_*]. \tag{3.1.4}$$

PROOF. We want to show that $\mathcal{E}_*$ is wandering for $\mathcal{U}$, i.e., that $\mathcal{U}^v\mathcal{E}_* \perp \mathcal{U}^w\mathcal{E}_*$, or equivalently, that

$$\mathcal{U}^{*w^\top}\mathcal{U}^v\mathcal{E} \perp \mathcal{E} \text{ for } v, w \in \mathcal{F}_d,\ v \neq w. \tag{3.1.5}$$

Since $\mathcal{U}$ is row unitary and $v \neq w$, it follows that $\mathcal{U}^{*w^\top}\mathcal{U}^v$ has one of the three forms:

$$\mathcal{U}^{*w^\top}\mathcal{U}^v = 0 \tag{1}$$

$$\mathcal{U}^{*w^\top}\mathcal{U}^v = \mathcal{U}^\alpha \text{ for some nonempty } \alpha \in \mathcal{F}_d \tag{2}$$

$$\mathcal{U}^{*w^\top}\mathcal{U}^v = \mathcal{U}^{*\beta} \text{ for some nonempty } \beta \in \mathcal{F}_d \tag{3}$$

In case (1), (3.1.5) is trivial. In case (2), write $\alpha = g_j\alpha'$. Since $\mathcal{E}_* \perp \mathcal{G}_*$ by definition, we have

$$\mathcal{U}^\alpha\mathcal{E}_* \perp \mathcal{U}^\alpha\mathcal{G}_*. \tag{3.1.6}$$

Since each $\mathcal{U}_k$ is an isometry and $\mathcal{U}_k^*\mathcal{G}_* \subset \mathcal{G}_*$, we have $\mathcal{U}_k\mathcal{G}_* \supset \mathcal{G}_*$, and hence $\mathcal{U}^\alpha\mathcal{G}_* \supset \mathcal{U}_j\mathcal{G}_*$. In particular, from (3.1.6) we see that

$$\mathcal{U}^\alpha\mathcal{E}_* \perp \mathcal{U}_j\mathcal{G}_*. \tag{3.1.7}$$

Moreover, as $\mathcal{U}^\alpha\mathcal{E}_* \subset \mathcal{U}_j\mathcal{K}$ and $\mathcal{U}_j\mathcal{K} \perp \mathcal{U}_k\mathcal{K} \supset \mathcal{U}_k\mathcal{G}_*$ for all $k \neq j$, we see that

$$\mathcal{U}^\alpha\mathcal{E}_* \perp \mathcal{U}_k\mathcal{G}_* \text{ for } k \neq j. \tag{3.1.8}$$

Combining (3.1.7) and (3.1.8) gives that $\mathcal{U}^\alpha\mathcal{E}_* \perp [\oplus_{k=1}^d\mathcal{U}_k\mathcal{G}_*] \supset \mathcal{E}_*$ and case (2) follows. In case (3), simply note from the row unitary property of $\mathcal{U}$ that

$$\mathcal{U}^{*\beta}\mathcal{E}_* \subset \mathcal{U}^{*\beta}[\oplus_{k=1}^d\mathcal{U}_k\mathcal{G}_*] \subset \mathcal{G}_*$$

where $\mathcal{G}_* \perp \mathcal{E}_*$ by definition. Hence $\mathcal{E}_*$ is wandering as asserted.

Set $\widetilde{\mathcal{G}}_{*0} = \mathcal{G}_* \oplus [\oplus_{w \in \mathcal{F}_d}\mathcal{U}^w\mathcal{E}_*]$. By construction $\mathcal{U}_j^* \colon \mathcal{E}_* \to \mathcal{G}_*$ and $\mathcal{U}_j \colon \mathcal{G}_* \to \mathcal{E}_* \oplus \mathcal{G}_*$ for each $j = 1, \dots, d$. It follows that $\widetilde{\mathcal{G}}_{*0}$ is reducing for each $\mathcal{U}_1, \dots, \mathcal{U}_d$. Since $\widetilde{\mathcal{G}}_*$ by definition is the smallest subspace of $\mathcal{K}$ reducing for each $\mathcal{U}_j$ and containing $\mathcal{E}_*$ and also contains $\mathcal{G}_*$ by hypothesis, it follows that necessarily $\widetilde{\mathcal{G}}_* = \widetilde{\mathcal{G}}_{*0}$. This completes the proof of Proposition 3.1.1. □

Another fundamental piece of the geometry of a Cuntz scattering system is given by the following Proposition.

PROPOSITION 3.1.2. *Let $\mathfrak{S}$ be a Cuntz scattering system* (3.1.1), *with scattering space $\mathcal{H} = \mathcal{K} \ominus [\mathcal{G}_* \oplus \mathcal{G}]$, incoming wandering subspace $\mathcal{E}_* = \left[\oplus_{j=1}^d \mathcal{U}_j \mathcal{G}_*\right] \ominus \mathcal{G}_*$ and outgoing wandering subspace $\mathcal{E} = \mathcal{G} \ominus \left[\oplus_{j=1}^d \mathcal{U}_j \mathcal{G}\right]$. Then*

$$\mathcal{H} \oplus \mathcal{E} = \mathcal{E}_* \oplus \begin{bmatrix} \mathcal{U}_1 & \dots & \mathcal{U}_d \end{bmatrix} \begin{bmatrix} \mathcal{H} \\ \vdots \\ \mathcal{H} \end{bmatrix}, \tag{3.1.9}$$

$$\begin{bmatrix} \mathcal{U}_1^* \\ \vdots \\ \mathcal{U}_d^* \end{bmatrix} (\mathcal{H} \oplus \mathcal{E}) = \begin{bmatrix} \mathcal{U}_1^* \\ \vdots \\ \mathcal{U}_d^* \end{bmatrix} \mathcal{E}_* \oplus \begin{bmatrix} \mathcal{H} \\ \vdots \\ \mathcal{H} \end{bmatrix}. \tag{3.1.10}$$

PROOF. The identity (3.1.10) follows immediately from (3.1.9) by multiplying both sides of (3.1.9) by $\begin{bmatrix} \mathcal{U}_1 & \dots & \mathcal{U}_d \end{bmatrix}^*$ and using the row unitary property of $(\mathcal{U}_1, \dots, \mathcal{U}_d)$.

We prove (3.1.9) as follows. First note that the orthogonal decomposition $\mathcal{H} \oplus \mathcal{E}$ is clear from $\mathcal{H} \perp \mathcal{G}$ and $\mathcal{E} \subset \mathcal{G}$. From $\mathcal{E}_* \subset \begin{bmatrix} \mathcal{U}_1 & \dots & \mathcal{U}_d \end{bmatrix} \left[\oplus_{j=1}^d \mathcal{G}_*\right]$, the known orthogonality $\mathcal{G}_* \perp \mathcal{H}$, and the row unitary property of $(\mathcal{U}_1, \dots, \mathcal{U}_d)$, the orthogonal decomposition $\mathcal{E}_* \oplus \begin{bmatrix} \mathcal{U}_1 & \dots & \mathcal{U}_d \end{bmatrix} \left[\oplus_{j=1}^d \mathcal{H}\right]$ is immediate as well. To prove the equality in (3.1.9), note that the orthogonal decomposition

$$\mathcal{K} = \mathcal{G}_* \oplus \mathcal{H} \oplus \mathcal{G}$$

trivially implies the orthogonal decomposition

$$\oplus_{j=1}^d \mathcal{K} = (\oplus_{j=1}^d \mathcal{G}_*) \oplus (\oplus_{j=1}^d \mathcal{H}) \oplus (\oplus_{j=1}^d \mathcal{G}).$$

The row unitary property of $\mathcal{U} = (\mathcal{U}_1, \dots, \mathcal{U}_d)$ then implies

$$\begin{aligned} \mathcal{K} &= \begin{bmatrix} \mathcal{U}_1 & \dots & \mathcal{U}_d \end{bmatrix} (\oplus_{j=1}^d \mathcal{K}) \\ &= \begin{bmatrix} \mathcal{U}_1 & \dots & \mathcal{U}_d \end{bmatrix} (\oplus_{j=1}^d \mathcal{G}_*) \oplus \begin{bmatrix} \mathcal{U}_1 & \dots & \mathcal{U}_d \end{bmatrix} (\oplus_{j=1}^d \mathcal{H}) \oplus \begin{bmatrix} \mathcal{U}_1 & \dots & \mathcal{U}_d \end{bmatrix} (\oplus_{j=1}^d \mathcal{G}) \\ &= \mathcal{G}_* \oplus \mathcal{E}_* \oplus \begin{bmatrix} \mathcal{U}_1 & \dots & \mathcal{U}_d \end{bmatrix} (\oplus_{j=1}^d \mathcal{H}) \oplus \begin{bmatrix} \mathcal{U}_1 & \dots & \mathcal{U}_d \end{bmatrix} (\oplus_{j=1}^d \mathcal{G}). \end{aligned} \tag{3.1.11}$$

On the other hand we have

$$\mathcal{K} = \mathcal{G}_* \oplus \mathcal{H} \oplus \mathcal{G} \tag{3.1.12}$$

$$= \mathcal{G}_* \oplus \mathcal{H} \oplus \mathcal{E} \oplus \begin{bmatrix} \mathcal{U}_1 & \dots & \mathcal{U}_d \end{bmatrix} (\oplus_{j=1}^d \mathcal{G}). \tag{3.1.13}$$

If we identify the two expressions for the orthogonal complement of

$$\mathcal{G}_* \oplus \begin{bmatrix} \mathcal{U}_1 & \dots & \mathcal{U}_d \end{bmatrix} (\oplus_{j=1}^d \mathcal{G})$$

obtained by computing in the two decompositions (3.1.11) and (3.1.13) for $\mathcal{K}$, we arrive at (3.1.9). □

In general, we shall say that the Cuntz scattering system $\mathfrak{S}$ is *minimal* if the span $\widetilde{\mathcal{G}}_* + \widetilde{\mathcal{G}}$ is dense in the ambient space $\mathcal{K}$. The subspace $\mathcal{H}$ defined by

$$\mathcal{H} = \mathcal{K} \ominus [\mathcal{G}_* \oplus \mathcal{G}]$$

will be called the *scattering space.*

As discussed in Chapter 2, the operators $W = [W_{v,w;\alpha,\beta}]_{v,w,\alpha,\beta\in\mathcal{F}_d}$, $W_* = [W_{*v,w;\alpha,\beta}]_{v,w,\alpha,\beta\in\mathcal{F}_d}$ and $W_{\mathcal{H}} = [W_{\mathcal{H},v,w;\alpha,\beta}]_{v,w,\alpha,\beta\in\mathcal{F}_d}$ defined by

$$W_{v,w;\alpha,\beta} = P_{\mathcal{E}}\mathcal{U}^w\mathcal{U}^{*v}\mathcal{U}^{\alpha^\top}\mathcal{U}^{*\beta^\top}|_{\mathcal{E}} \tag{3.1.14}$$

$$W_{*v,w;\alpha,\beta} = P_{\mathcal{E}_*}\mathcal{U}^w\mathcal{U}^{*v}\mathcal{U}^{\alpha^\top}\mathcal{U}^{*\beta^\top}|_{\mathcal{E}_*}. \tag{3.1.15}$$

$$W_{\mathcal{H},v,w;\alpha,\beta} = P_{\mathcal{H}}\mathcal{U}^w\mathcal{U}^{*v}\mathcal{U}^{\alpha^\top}\mathcal{U}^{*\beta^\top}|_{\mathcal{H}} \tag{3.1.16}$$

are all positive-semidefinite Cuntz weights with associated factorizations $W = \Phi\Phi^{[*]}$, $W_* = \Phi_*\Phi_*^{[*]}$ and $W_{\mathcal{H}} = \Phi_{\mathcal{H}}\Phi_{\mathcal{H}}^{[*]}$, where the operators Φ, Φ_* and $\Phi_{\mathcal{H}}$ given by

$$\begin{aligned} \Phi\colon k &\mapsto \sum_{v,w\in\mathcal{F}_d} (P_{\mathcal{E}}\mathcal{U}^w\mathcal{U}^{*v}k)z^v\zeta^w, \\ \Phi_*\colon k &\mapsto \sum_{v,w\in\mathcal{F}_d} (P_{\mathcal{E}_*}\mathcal{U}^w\mathcal{U}^{*v}k)z^v\zeta^w \text{ and} \\ \Phi_{\mathcal{H}}\colon k &\mapsto \sum_{v,w\in\mathcal{F}_d} (P_{\mathcal{H}}\mathcal{U}^w\mathcal{U}^{*v}k)z^v\zeta^w \end{aligned} \tag{3.1.17}$$

are coisometries from $\mathcal{K}$ onto $\mathcal{L}_W$, $\mathcal{L}_{W_*}$ and $\mathcal{L}_{W_{\mathcal{H}}}$, respectively. We shall be primarily interested only in Φ and Φ_* as the associated auxiliary subspaces $\mathcal{E}$ and $\mathcal{E}_*$ are wandering subspaces for $\mathcal{U}$ for these cases. For this reason it follows that W and W_* have the additional property that

$$\begin{aligned} W_{v,\emptyset;\alpha,\emptyset} &= \delta_{v,\alpha}, \\ W_{*v,\emptyset;\alpha,\emptyset} &= \delta_{v,\alpha}, \end{aligned}$$

i.e., W and W_* are Cuntz-weight extensions of the identity, from which we get

$$\begin{aligned} \mathcal{H}_W &= WL^2(\mathcal{F}_d\times\{\emptyset\},\mathcal{E}) \subsetneq \mathcal{L}_W, \\ \mathcal{H}_{W_*} &= W_*L^2(\mathcal{F}_d\times\{\emptyset\},\mathcal{E}_*) \subsetneq \mathcal{L}_{W_*}. \end{aligned} \tag{3.1.18}$$

From the discussion in Chapter 2 (see [**BaV04**] for complete details) we also have the intertwining relations

$$\Phi\mathcal{U}_j = \mathcal{U}_{W,j}\Phi, \qquad \Phi_*\mathcal{U}_j = \mathcal{U}_{W_*,j}\Phi_*, \qquad \Phi_{\mathcal{H}}\mathcal{U}_j = \mathcal{U}_{W_{\mathcal{H}},j}\Phi_{\mathcal{H}} \tag{3.1.19}$$

as well as

$$\Phi\mathcal{U}_j^* = \mathcal{U}_{W,j}^*\Phi, \qquad \Phi_*\mathcal{U}_j^* = \mathcal{U}_{W_*,j}^*\Phi_*, \qquad \Phi_{\mathcal{H}}\mathcal{U}_j^* = \mathcal{U}_{\mathcal{H},j}^*\Phi_{\mathcal{H}} \tag{3.1.20}$$

for $j = 1,\dots,d$. In addition, from the construction and the decomposition (3.1.4), we see that

$$\Phi(\mathcal{G}) = \mathcal{H}_W, \qquad \Phi_*(\mathcal{G}_*) = \mathcal{H}_{W_*}^\perp.$$

and the spaces $\mathcal{L}_W$, $\mathcal{L}_{W_*}$ and $\mathcal{L}_{W_{\mathcal{H}}}$ are characterized as

$$\begin{aligned} \mathcal{L}_W &= \left\{\sum u(v,w)z^v\zeta^w\colon u(v,w) = P_{\mathcal{E}}\mathcal{U}^w\mathcal{U}^{*v}k \text{ for some } k\in\mathcal{K}\right\}, \\ \mathcal{L}_{W_*} &= \left\{\sum y(v,w)z^v\zeta^w\colon y(v,w) = P_{\mathcal{E}_*}\mathcal{U}^w\mathcal{U}^{*v}k \text{ for some } k\in\mathcal{K}\right\}, \\ \mathcal{L}_{W_{\mathcal{H}}} &= \left\{\sum x(v,w)z^v\zeta^w\colon x(v,w) = P_{\mathcal{H}}\mathcal{U}^w\mathcal{U}^{*v}k \text{ for some } k\in\mathcal{K}\right\}. \end{aligned} \tag{3.1.21}$$

3.2. The scattering function and models for Cuntz scattering systems

Given a Cuntz scattering system $\mathfrak{S}$, we define the scattering operator $\mathbf{S}\colon \widetilde{\mathcal{G}} \to \widetilde{\mathcal{G}}_*$ by

$$\mathbf{S} = P_{\widetilde{\mathcal{G}}_*}|_{\mathcal{G}}.$$

Since $\mathbf{S}$ is the restriction of an orthogonal projection, we see that $\|\mathbf{S}\| \leq 1$. Since $\widetilde{\mathcal{G}}$ and $\widetilde{\mathcal{G}}_*$ are reducing for each $\mathcal{U}_j$ we see immediately that $\mathbf{S}$ satisfies the intertwining relations

$$\mathbf{S}\mathcal{U}_j|_{\widetilde{\mathcal{G}}} = \mathcal{U}_j\mathbf{S}, \tag{3.2.1}$$

$$\mathbf{S}\mathcal{U}_j^*|_{\widetilde{\mathcal{G}}} = \mathcal{U}_j^*\mathbf{S} \tag{3.2.2}$$

for each $j = 1, \dots, d$. The assumed orthogonality between $\mathcal{G}$ and $\mathcal{G}_*$ implies that $\mathbf{S}\colon \mathcal{G} \to [\widetilde{\mathcal{G}}_* \ominus \mathcal{G}_*]$, or

$$\mathbf{S}\colon [\oplus_{v\in\mathcal{F}_d}\mathcal{U}^v\mathcal{E}] \to [\oplus_{v\in\mathcal{F}_d}\mathcal{U}^v\mathcal{E}_*]. \tag{3.2.3}$$

To put this in more concrete form, we introduce functional models for $\widetilde{\mathcal{G}}_*$ and $\widetilde{\mathcal{G}}$ via Fourier representations as in the classical case.

We define the *scattering function* S by

$$S = \Phi_*\Phi^*\colon \mathcal{L}_W \to \mathcal{L}_{W_*}.$$

Note that since the initial space for Φ is $\widetilde{\mathcal{G}}$ and the initial space for Φ_* is $\widetilde{\mathcal{G}}_*$, we could have equivalently defined S as

$$S = \Phi_*(P_{\widetilde{\mathcal{G}}_*}|_{\widetilde{\mathcal{G}}})\Phi^* = \Phi_*\mathbf{S}\Phi^*\colon \mathcal{L}_W \to \mathcal{L}_{W_*}.$$

From the intertwining conditions (3.1.19) and (3.1.20) (or from (3.2.1) and (3.2.2)) we see that S satisfies the intertwining relations

$$\begin{aligned} S\mathcal{U}_{W,j} &= \mathcal{U}_{W_*,j}S, \\ S\mathcal{U}_{W,j}^* &= \mathcal{U}_{W_*,j}^*S \text{ for each } j = 1, \dots, d. \end{aligned}$$

As S is the composition of partial isometries, we see immediately that S is a contraction operator ($\|S\| \leq 1$). From the causality condition (3.1.2) (or from (3.2.3)) we see that

$$S|_{\mathcal{H}_W}\colon \mathcal{H}_W \to \mathcal{H}_{W_*}$$

and hence S is in the Schur class $\mathcal{S}(W, W_*)$ of contractive, analytic multipliers between $\mathcal{L}_W$ and $\mathcal{L}_{W_*}$.

Conversely (see Theorems 3.2.1, 3.2.2 and 3.2.3 below), given any two positive semidefinite Cuntz weights W and W_* and a Schur-class multiplier $S \in \mathcal{S}(W, W_*)$, there is a Cuntz scattering system $\mathfrak{S}^S$ having scattering function (after trivial modifications) equal to S. Furthermore, if $\mathfrak{S}$ is any minimal Cuntz scattering system with scattering function equal to S, then $\mathfrak{S}$ is unitarily equivalent to $\mathfrak{S}^S$, and hence the scattering function S is a complete unitary invariant for minimal Cuntz scattering systems, just as in the classical case $d = 1$.

First we make precise the notion of "trivial identifications". Suppose that $S \in \mathcal{S}(W, W_*)$ and $S' \in \mathcal{S}(W', W'_*)$ are two Schur-class multipliers, where W, W_*, W', W'_* are positive-semidefinite Cuntz weights on the Hilbert spaces $\mathcal{E}$, $\mathcal{E}_*$, $\mathcal{E}'$ and $\mathcal{E}'_*$ respectively. Suppose that there are unitary identification maps $i\colon \mathcal{E} \to \mathcal{E}'$ and $i_*\colon \mathcal{E}_* \to \mathcal{E}'_*$ such that

$$W_{v,w;\alpha,\beta} = i^*W'_{v,w;\alpha,\beta}i, \qquad W_{*v,w;\alpha,\beta} = i_*^*W'_{*v,w;\alpha,\beta}i_*.$$

for all words $v, w \in \mathcal{F}_d$. Suppose in addition that $i_* f_{v,w} = f'_{v,w}$ for all $v, w \in \mathcal{F}_d$ whenever

$$S'(iez^\alpha\zeta^\beta) = \sum_{v,w} f'_{v,w} z^v \zeta^w, \qquad S(ez^\alpha\zeta^\beta) = \sum_{v,w} f_{v,w} z^v \zeta^w$$

for some $\alpha, \beta \in \mathcal{F}_d$ and $e \in \mathcal{E}$. Under these circumstances we say that the Schur-class multipliers S and S' *coincide*.

One way to construct such an $\mathfrak{S}$ is via the Pavlov model, which we now define. The Pavlov model is a scattering system $\mathfrak{S}_P$ constructed explicitly from the information encoded in a given Schur-class multiplier $S \in \mathcal{S}(W, W_*)$; for simplicity, we suppress the Schur-class multiplier S from the notation. The Pavlov model $\mathfrak{S}_P$ is the Cuntz scattering system

$$\mathfrak{S}_P = (\mathcal{U}_P = (\mathcal{U}_{P,1}, \dots, \mathcal{U}_{P,d}); \mathcal{K}_P, \mathcal{G}_P, \mathcal{G}_{P*}) \tag{3.2.4}$$

defined as follows. The ambient space $\mathcal{K}_P$ is defined by

$$\mathcal{K}_P = \text{ closure of equivalence classes of elements in } \begin{bmatrix} \mathcal{L}_{W_*} \\ \mathcal{L}_W \end{bmatrix}$$

with $\mathcal{K}_P$-inner product given densely by

$$\left\langle \begin{bmatrix} f \\ g \end{bmatrix}, \begin{bmatrix} f' \\ g' \end{bmatrix} \right\rangle_{\mathcal{K}_P} = \left\langle \begin{bmatrix} I & S \\ S^* & I \end{bmatrix} \begin{bmatrix} f \\ g \end{bmatrix}, \begin{bmatrix} f' \\ g' \end{bmatrix} \right\rangle_{\mathcal{L}_{W_*} \oplus \mathcal{L}_W},$$

with outgoing and incoming spaces given by

$$\mathcal{G}_P = \begin{bmatrix} 0 \\ \mathcal{H}_W \end{bmatrix}, \qquad \mathcal{G}_{P*} = \begin{bmatrix} \mathcal{H}_{W_*}^\perp \\ 0 \end{bmatrix},$$

and with row-unitary d-tuple $\mathcal{U}_P = (\mathcal{U}_{P,1}, \dots, \mathcal{U}_{P,d})$ given by

$$\mathcal{U}_{P,j} = \begin{bmatrix} \mathcal{U}_{W_*,j} & 0 \\ 0 & \mathcal{U}_{W,j} \end{bmatrix}.$$

Then the following is the basic result concerning the Pavlov model.

THEOREM 3.2.1. *Let S be a Schur-class multiplier in $\mathcal{S}(W, W_*)$ for two given positive semidefinite Cuntz-weight extensions of the identity W and W_* on auxiliary Hilbert spaces $\mathcal{E}$ and $\mathcal{E}_*$ respectively. Then the system $\mathfrak{S}_P$ defined by (3.2.4) is a minimal Cuntz scattering system with outgoing and incoming wandering subspaces given by*

$$\mathcal{E}_P = \begin{bmatrix} 0 \\ W\mathcal{E} \end{bmatrix}, \qquad \mathcal{E}_{P*} = \begin{bmatrix} W_*\mathcal{E}_* \\ 0 \end{bmatrix},$$

with scattering function S_P coinciding with S via the identification operators

$$i_P \colon \mathcal{E} \to \mathcal{E}_P \text{ and } i_{P*} \colon \mathcal{E}_* \to \mathcal{E}_{P*}$$

given by

$$i_P \colon e \mapsto \begin{bmatrix} 0 \\ W[e] \end{bmatrix}, \qquad i_{P*} \colon e_* \mapsto \begin{bmatrix} W_*[e_*] \\ 0 \end{bmatrix}.$$

Moreover, any other minimal Cuntz scattering system (3.1.1) $\mathfrak{S}$ with scattering function equal to S is unitarily equivalent to the Pavlov-model scattering system $\mathfrak{S}_P$ associated with S via the identification map $\mathcal{I}_P \colon \mathcal{K} \to \mathcal{K}_P$ given by

$$\mathcal{I}_P \colon \Phi_*^* f + \Phi^* g \mapsto \begin{bmatrix} f \\ g \end{bmatrix} \text{ for } f \in \mathcal{L}_{W_*} \text{ and } g \in \mathcal{L}_W.$$

PROOF. Since W and W_* are Cuntz-weight extensions of the identity, it is easily checked that $\mathfrak{S}_P$ is a Cuntz scattering system with outgoing wandering subspace $\mathcal{E}_P = \begin{bmatrix} 0 \\ W\mathcal{E} \end{bmatrix}$ and with incoming wandering subspace $\mathcal{E}_{P*} = \begin{bmatrix} W_*\mathcal{E}_* \\ 0 \end{bmatrix}$. Then the associated spun-out outgoing and incoming subspaces work out to be

$$\widetilde{\mathcal{G}}_P = \begin{bmatrix} 0 \\ \mathcal{L}_W \end{bmatrix}, \qquad \widetilde{\mathcal{G}}_{P*} = \begin{bmatrix} \mathcal{L}_{W_*} \\ 0 \end{bmatrix}$$

and the scattering function S_P coincides with the original Schur-class multiplier S via the identification maps $i\colon \mathcal{E} \to \mathcal{E}_P$ and $i_*\colon \mathcal{E}_* \to \mathcal{E}_{*P}$ given by

$$i_P\colon e \mapsto \begin{bmatrix} 0 \\ W[e] \end{bmatrix}, \qquad i_{P*}\colon e_* \mapsto \begin{bmatrix} W_*[e_*] \\ 0 \end{bmatrix}.$$

Finally, if $\mathfrak{S}$ is any Cuntz scattering system with scattering function S, one can easily check that the map $\mathcal{I}_P\colon \mathcal{K} \to \mathcal{K}_P$ given by

$$\mathcal{I}_P\colon \Phi_*^* f + \Phi^* g \mapsto \begin{bmatrix} f \\ g \end{bmatrix} \quad \text{for } f \in \mathcal{L}_{W_*} \text{ and } g \in \mathcal{L}_W$$

extends to a coisometry from $\mathcal{K}$ onto $\mathcal{K}_P$ with kernel (if any) equal to $[\widetilde{\mathcal{G}}_* + \widetilde{\mathcal{G}}]^\perp$ which implements a unitary equivalence between the Cuntz scattering system equal to the minimal part of $\mathfrak{S}$ and the scattering system $\mathfrak{S}_P$. This concludes the proof of Theorem 3.2.1. $\square$

The de Branges-Rovnyak model for a given Schur-class multiplier $S \in \mathcal{S}(W, W_*)$

$$\mathfrak{S}_{dBR} = (\mathcal{U}_{dBR} = (\mathcal{U}_{dBR,1}, \dots, \mathcal{U}_{dBR,d}); \mathcal{K}_{dBR}, \mathcal{G}_{dBR}, \mathcal{G}_{dBR*}) \tag{3.2.5}$$

is defined as follows. The ambient space is

$$\mathcal{K}_{dBR} = \operatorname{im} \begin{bmatrix} I & S \\ S^* & I \end{bmatrix}^{1/2} \subset \begin{bmatrix} \mathcal{L}_{W_*} \\ \mathcal{L}_W \end{bmatrix}$$

with lifted inner product

$$\left\langle \begin{bmatrix} I & S \\ S^* & I \end{bmatrix}^{1/2} \begin{bmatrix} f \\ g \end{bmatrix}, \begin{bmatrix} I & S \\ S^* & I \end{bmatrix}^{1/2} \begin{bmatrix} f' \\ g' \end{bmatrix} \right\rangle_{\mathcal{K}_{dBR}} = \left\langle Q \begin{bmatrix} f \\ g \end{bmatrix}, \begin{bmatrix} f' \\ g' \end{bmatrix} \right\rangle_{\mathcal{L}_{W_*} \oplus \mathcal{L}_W}$$

where Q is the orthogonal projection onto the orthogonal complement of the kernel of $\begin{bmatrix} I & S \\ S^* & I \end{bmatrix}$, with outgoing and incoming subspaces equal to

$$\mathcal{G}_{dBR} = \begin{bmatrix} S \\ I \end{bmatrix} \mathcal{H}_W, \qquad \mathcal{G}_{dBR*} = \begin{bmatrix} I \\ S^* \end{bmatrix} \mathcal{H}_{W_*}^\perp$$

and row unitary d-tuple $\mathcal{U}_{dBR} = (\mathcal{U}_{dBR,1}, \dots, \mathcal{U}_{dBR,d})$ on $\mathcal{K}_{dBR}$ given by

$$\mathcal{U}_{dBR,j} = \begin{bmatrix} \mathcal{U}_{W_*,j} & 0 \\ 0 & \mathcal{U}_{W,j} \end{bmatrix}.$$

Then the analogue of Theorem 3.2.1 for the de Branges-Rovnyak model is the following.

THEOREM 3.2.2. *Let S be a Schur-class multiplier in $\mathcal{S}(W, W_*)$ for two given positive semidefinite Cuntz-weight extensions of the identity W and W_* on auxiliary Hilbert spaces $\mathcal{E}$ and $\mathcal{E}_*$ respectively. Then the system $\mathfrak{S}_{dBR}$ defined by* (3.2.5) *is a* *minimal Cuntz scattering system with outgoing and incoming wandering subspaces given by*

$$\mathcal{E}_{dBR} = \begin{bmatrix} S \\ I \end{bmatrix} W\mathcal{E}, \qquad \mathcal{E}_{dBR*} = \begin{bmatrix} I \\ S^* \end{bmatrix} W_*\mathcal{E}_*,$$

with scattering function S_{dBR} *coinciding with* S *via the identification operators* $i_{dBR}\colon \mathcal{E} \to \mathcal{E}_{dBR}$ *and* $i_{dBR*}\colon \mathcal{E}_* \to \mathcal{E}_{dBR*}$ *given by*

$$i_{dBR}\colon e \mapsto \begin{bmatrix} S \\ I \end{bmatrix} W[e], \qquad i_{dBR*}\colon e_* \mapsto \begin{bmatrix} I \\ S \end{bmatrix} W_*[e_*]$$

Moreover, any other minimal Cuntz scattering system $\mathfrak{S}$ (3.1.1) *with scattering function equal to* S *is unitarily equivalent to the de Branges-Rovnyak-model scattering system* $\mathfrak{S}_{dBR}$ *associated with* S *via the identification map* $\mathcal{I}_{dBR}\colon \mathcal{K} \to \mathcal{K}_{dBR}$ *given by*

$$\mathcal{I}_{dBR}\colon \Phi_*^* f + \Phi^* g \mapsto \begin{bmatrix} I & S \\ S^* & I \end{bmatrix} \begin{bmatrix} f \\ g \end{bmatrix}.$$

The Sz.-Nagy-Foiaş Cuntz scattering model for a given Schur-class multiplier $S \in \mathcal{S}(W, W_*)$ is defined to be the system

$$\mathfrak{S}_{NF} = (\mathcal{U}_{NF} = (\mathcal{U}_{NF,1}, \dots, \mathcal{U}_{NF,d}); \mathcal{K}_{NF}, \mathcal{G}_{NF}, \mathcal{G}_{NF*}) \tag{3.2.6}$$

with ambient space

$$\mathcal{K}_{NF} = \begin{bmatrix} \mathcal{L}_{W_*} \\ \text{clos. } D_S\mathcal{L}_W \end{bmatrix} \subset \begin{bmatrix} \mathcal{L}_{W_*} \\ \mathcal{L}_W \end{bmatrix}$$

where D_S is the defect operator $D_S = (I - S^*S)^{1/2}$ and the inner product is taken to be the inner product inherited from $\mathcal{L}_{W_*} \oplus \mathcal{L}_W$, with outgoing and incoming subspaces equal to

$$\mathcal{G}_{NF} = \begin{bmatrix} S \\ D_S \end{bmatrix} \mathcal{H}_W, \qquad \mathcal{G}_{NF*} = \begin{bmatrix} \mathcal{H}_{W*}^\perp \\ 0 \end{bmatrix}$$

and row unitary d-tuple $\mathcal{U}_{NF} = (\mathcal{U}_{NF,1}, \dots, \mathcal{U}_{NF,d})$ on $\mathcal{K}_{NF}$ given by

$$\mathcal{U}_{NF,j} = \begin{bmatrix} \mathcal{U}_{W_*,j} & 0 \\ 0 & \mathcal{U}_{W,j} \end{bmatrix}.$$

Then the analogue of Theorem 3.2.1 for the Sz.-Nagy-Foiaş model is the following.

THEOREM 3.2.3. *Let* S *be a Schur-class multiplier in* $\mathcal{S}(W, W_*)$ *for two given positive semidefinite Cuntz-weight extensions of the identity* W *and* W_* *on auxiliary Hilbert spaces* $\mathcal{E}$ *and* $\mathcal{E}_*$ *respectively. Then the system* $\mathfrak{S}_{NF}$ *defined by* (3.2.6) *is a minimal Cuntz scattering system with outgoing and incoming wandering subspaces given by*

$$\mathcal{E}_{NF} = \begin{bmatrix} S \\ D_S \end{bmatrix} W\mathcal{E}, \qquad \mathcal{E}_{NF*} = \begin{bmatrix} W_*\mathcal{E}_* \\ 0 \end{bmatrix},$$

with scattering function S_{NF} *coinciding with* S *via the identification operators* $i_{NF}\colon \mathcal{E} \to \mathcal{E}_{NF}$ *and* $i_{NF*}\colon \mathcal{E}_* \to \mathcal{E}_{NF*}$ *given by*

$$i_{NF}\colon e \mapsto \begin{bmatrix} S \\ D_S \end{bmatrix} W[e], \qquad i_{NF*}\colon e_* \mapsto \begin{bmatrix} W_*[e_*] \\ 0 \end{bmatrix}$$

Moreover, any other minimal Cuntz scattering system $\mathfrak{S}$ (3.1.1) *with scattering function equal to* S *is unitarily equivalent to the Sz.-Nagy-Foiaş-model scattering system* $\mathfrak{S}_{NF}$ *associated with* S *via the identification map* $\mathcal{I}_{NF}\colon \mathcal{K} \to \mathcal{K}_{NF}$ *given by*

$$\mathcal{I}_{NF}\colon \Phi_*^* f + \Phi^* g \mapsto \begin{bmatrix} I & S \\ 0 & D_S \end{bmatrix} \begin{bmatrix} f \\ g \end{bmatrix}.$$

As a corollary of any one of these models, we have the following result.

COROLLARY 3.2.4. *The scattering function* $S \in \mathcal{S}(W, W_*)$ *is a complete unitary invariant for the category of minimal Cuntz scattering systems. More precisely, two minimal Cuntz scattering systems* $\mathfrak{S}$ *and* $\mathfrak{S}'$ *are unitarily equivalent if and only if their associated scattering functions* S *and* S' *coincide.*

PROOF. By tracing through the definition of unitary equivalence of scattering systems, it is easily seen that unitary equivalence of scattering systems $\mathfrak{S}$ and $\mathfrak{S}'$ forces the associated scattering functions S and S' to coincide. Conversely, suppose that the scattering functions S and S' for the minimal scattering systems $\mathfrak{S}$ and $\mathfrak{S}'$ coincide. Then $\mathfrak{S}$ is unitarily equivalent to the Sz.-Nagy-Foiaş model $\mathfrak{S}_{NF}(S)$ based on S while $\mathfrak{S}'$ is unitarily equivalent to the Sz.-Nagy-Foiaş model $\mathfrak{S}_{NF}(S')$ based on S'. But it is easily checked at the model level that coincidence of characteristic functions S and S' forces unitary equivalence of the associated scattering systems $\mathfrak{S}_{NF}(S)$ and $\mathfrak{S}_{NF}(S')$. Hence the original scattering systems $\mathfrak{S}$ and $\mathfrak{S}'$ are unitarily equivalent as asserted. □

REMARK 3.2.5. By using any one of the models given above (Pavlov, de Branges-Rovnyak or Sz.-Nagy-Foiaş), we see that a complete unitary invariant for a minimal Cuntz scattering system is a contractive intertwining map $S \in \mathcal{S}(W, W_*)$ for some Cuntz-weight extensions of the identity W, W_*. By Proposition 2.4.1 any such S has the form L_T^{W,W_*} for a multiplier $T(z) = \sum_{v \in \mathcal{F}_d} T_v z^v$ with $\|M_T\| \le 1$. Conversely, given a triple (T, W, W_*), the associated operator $S = L_T^{W,W_*}$ is in $\mathcal{S}(W, W_*)$ if and only if (T, W, W_*) is admissible in the sense that one of the supplementary conditions in Theorem 2.4.2 (for the Cuntz-weight case) is satisfied. As explained in Theorem 2.4.5, another choice of parameters for describing an admissible triple (T, W, W_*) is (T, L, W_*), where L is a Cuntz-weight extension of the Cuntz-Toeplitz operator $I - (M_T)^* M_T$, or equivalently, (T, X, W_*) where $X(z, \zeta)$ is a positive symbol of the form (2.4.23) with associated defect symbol $X'(z, \zeta)$ (see (2.4.24)) equal to 0. Note that Cuntz-weight extensions L of $I - (M_T)^* M_T$ in turn can be parametrized in terms of Cuntz weight extensions of an identity by using the maximal factorable minorant for $I - (M_T)^* M_T$, as explained in Theorem 2.4.7. As we shall see, the parametrization (T, L, W_*) or (T, X, W_*) is more convenient for separating the role of the incoming Cuntz weight W_* from the role of the associated Cuntz unitary colligation U to be discussed in the next section; indeed, (T, L) or (T, X) form a set of complete unitary invariants for the "forward part" of a minimal Cuntz scattering system to be discussed in Chapter 5.

3.3. The colligation associated with a Cuntz scattering system

Let $\mathfrak{S} = (\mathcal{U} = (\mathcal{U}_1, \dots, \mathcal{U}_d); \mathcal{K}, \mathcal{G}, \mathcal{G}_*)$ be a Cuntz scattering system, with associated outgoing wandering subspace $\mathcal{E} = \mathcal{G} \ominus [\oplus_{j=1}^d \mathcal{U}_j \mathcal{G}]$, incoming wandering subspace $\mathcal{E}_* = [\oplus_{j=1}^d \mathcal{U}_j \mathcal{G}_*] \ominus \mathcal{G}_*$ and scattering space $\mathcal{H} = \mathcal{K} \ominus [\mathcal{G}_* \oplus \mathcal{G}]$ as in Chapter 3. In general by a (noncommutative, d-variable) colligation, we mean an operator of the form

$$U = \begin{bmatrix} A & B \\ C & D \end{bmatrix} = \begin{bmatrix} A_1 & B_1 \\ \vdots & \vdots \\ A_d & B_d \\ C & D \end{bmatrix} : \begin{bmatrix} \mathcal{H} \\ \mathcal{E} \end{bmatrix} \to \begin{bmatrix} \oplus_{j=1}^d \mathcal{H} \\ \mathcal{E}_* \end{bmatrix} \tag{3.3.1}$$

for some HIlbert spaces $\mathcal{H}$ (the *state space*), $\mathcal{E}$ (the *input space*) and $\mathcal{E}_*$ (the *output space*). In case U is unitary as an operator from $\mathcal{H} \oplus \mathcal{E}$ to $[\oplus_{j=1}^d \mathcal{H}] \oplus \mathcal{E}_*$, we say that U is a *unitary colligation*. Such objects will be discussed systematically in Chapter 4. Given a Cuntz scattering system $\mathfrak{S}$, we associate a colligation $U(\mathfrak{S})$ (3.3.1) by defining

$$\begin{aligned} \begin{bmatrix} A_j & B_j \end{bmatrix} &= P_{\mathcal{H}} \mathcal{U}_j^* |_{\mathcal{H} \oplus \mathcal{E}}, \\ \begin{bmatrix} C & D \end{bmatrix} &= P_{\mathcal{E}_*} |_{\mathcal{H} \oplus \mathcal{E}}. \end{aligned} \tag{3.3.2}$$

The first order of business is to prove the following.

THEOREM 3.3.1. *The colligation $U(\mathfrak{S})$ associated with a Cuntz scattering system as in* (3.3.2) *is unitary.*

PROOF. To show that $U(\mathfrak{S})$ is isometric, we must show that

$$\sum_{j=1}^d \|A_j h + B_j e\|^2 + \|Ch + De\|^2 = \|h\|^2 + \|e\|^2 \tag{3.3.3}$$

for all $h \in \mathcal{H}$ and $e \in \mathcal{E}$. From the definition (3.3.2), we see that (3.3.3) is equivalent to

$$\sum_{j=1}^d \|P_{\mathcal{H}} \mathcal{U}_j^* (h+e)\|^2 + \|P_{\mathcal{E}_*}(h+e)\|^2 = \|h\|^2 + \|e\|^2. \tag{3.3.4}$$

By Proposition 3.1.2 we know (3.1.9). Hence, for $h + e \in \mathcal{H} \oplus \mathcal{E}$, we have

$$\begin{aligned} \|h\|^2 + \|e\|^2 &= \|h+e\|^2 \\ &= \|P_{\mathcal{E}_*}(h+e)\|^2 + \left\| P_{[\mathcal{U}_1 \ \dots \ \mathcal{U}_d][\oplus_{j=1}^d \mathcal{H}]}(h+e) \right\|^2 \end{aligned} \tag{3.3.5}$$

Since $\begin{bmatrix} \mathcal{U}_1 & \dots & \mathcal{U}_d \end{bmatrix}$ is unitary,

$$P_{[\mathcal{U}_1 \ \dots \ \mathcal{U}_d][\oplus_{j=1}^d \mathcal{H}]} = \begin{bmatrix} \mathcal{U}_1 & \dots & \mathcal{U}_d \end{bmatrix} P_{[\oplus_{j=1}^d \mathcal{H}]} \begin{bmatrix} \mathcal{U}_1 & \dots & \mathcal{U}_d \end{bmatrix}^*.$$

Thus

$$\begin{aligned} \|P_{[\mathcal{U}_1 \ \dots \ \mathcal{U}_d][\oplus_{j=1}^d \mathcal{H}]}(h+e)\|^2 &= \| \begin{bmatrix} \mathcal{U}_1 & \dots & \mathcal{U}_d \end{bmatrix} P_{[\oplus_{j=1}^d \mathcal{H}]} \begin{bmatrix} \mathcal{U}_1 & \dots & \mathcal{U}_d \end{bmatrix}^* (h+e)\|^2 \\ &= \|P_{[\oplus_{j=1}^d \mathcal{H}]} \begin{bmatrix} \mathcal{U}_1 & \dots & \mathcal{U}_d \end{bmatrix}^* (h+e)\|^2 \\ &= \sum_{j=1}^d \|P_{\mathcal{H}} \mathcal{U}_j^* (h+e)\|^2 \end{aligned}$$

Upon combining this with (3.3.5) we get (3.3.4) as wanted.

To show that $U = U(\mathfrak{S})$ is unitary, it suffices now to show that U is surjective, i.e., given $h_1, \dots, h_d \in \mathcal{H}$ and $e_* \in \mathcal{E}_*$, we must be able to solve for an $h \in \mathcal{H}$ and an $e \in \mathcal{E}$ so that

$$\begin{bmatrix} A_1 & B_1 \\ \vdots & \vdots \\ A_d & B_d \\ C & D \end{bmatrix} \begin{bmatrix} h \\ e \end{bmatrix} = \begin{bmatrix} h_1 \\ \vdots \\ h_d \\ e_* \end{bmatrix},$$

or, equivalently,

$$P_{\mathcal{H}}\mathcal{U}_j^*(h+e) = h_j \text{ for } j = 1, \dots, d \text{ and } P_{\mathcal{E}_*}(h+e) = e_*. \tag{3.3.6}$$

From (3.1.10) we know that we can solve for $h \in \mathcal{H}$ and $e \in \mathcal{E}$ so that

$$\begin{bmatrix} \mathcal{U}_1^* \\ \vdots \\ \mathcal{U}_d^* \end{bmatrix} (h+e) = \begin{bmatrix} \mathcal{U}_1^* \\ \vdots \\ \mathcal{U}_d^* \end{bmatrix} e_* + \begin{bmatrix} h_1 \\ \vdots \\ h_d \end{bmatrix}.$$

From this identity we read off

$$P_{\mathcal{H}}\mathcal{U}_j^*(h+e) = h_j \text{ for } j = 1, \dots, d,$$
$$P_{[\mathcal{U}_1 \ \dots \ \mathcal{U}_d]^*\mathcal{E}_*} \begin{bmatrix} \mathcal{U}_1 & \dots & \mathcal{U}_d \end{bmatrix}^* (h+e) = \begin{bmatrix} \mathcal{U}_1 & \dots & \mathcal{U}_d \end{bmatrix}^* e_*. \tag{3.3.7}$$

Thus the pair (h, e) solves the first d equations in (3.3.6), and it remains only to verify the last equation.

Since $\begin{bmatrix} \mathcal{U}_1 & \dots & \mathcal{U}_d \end{bmatrix}$ is unitary, we have the congruence of projections

$$P_{[\mathcal{U}_1 \ \dots \ \mathcal{U}_d]^*\mathcal{E}_*} = \begin{bmatrix} \mathcal{U}_1 & \dots & \mathcal{U}_d \end{bmatrix}^* P_{\mathcal{E}_*} \begin{bmatrix} \mathcal{U}_1 & \dots & \mathcal{U}_d \end{bmatrix}, \tag{3.3.8}$$

and hence

$$P_{[\mathcal{U}_1 \ \dots \ \mathcal{U}_d]^*\mathcal{E}_*} \begin{bmatrix} \mathcal{U}_1 & \dots & \mathcal{U}_d \end{bmatrix}^* (h+e) = \begin{bmatrix} \mathcal{U}_1 & \dots & \mathcal{U}_d \end{bmatrix}^* P_{\mathcal{E}_*}(h+e)$$

and the last of equations (3.3.7) becomes

$$\begin{bmatrix} \mathcal{U}_1 & \dots & \mathcal{U}_d \end{bmatrix}^* P_{\mathcal{E}_*}(h+e) = \begin{bmatrix} \mathcal{U}_1 & \dots & \mathcal{U}_d \end{bmatrix}^* e_*. \tag{3.3.9}$$

Cancelling off the unitary factor $\begin{bmatrix} \mathcal{U}_1 & \dots & \mathcal{U}_d \end{bmatrix}^*$ in (3.3.9) leaves us with the last of equations (3.3.6) as wanted. This completes the proof of Theorem 3.3.1. □

We shall need the following formula for the incoming wandering subspace $\mathcal{E}_*$ in terms of the colligation operators A_j, B_j, C, D.

PROPOSITION 3.3.2. *Let $\mathfrak{S}$ be a Cuntz scattering system with associated unitary colligation $U(\mathfrak{S})$ as in (3.3.2). Then the incoming wandering subspace $\mathcal{E}_*$ can be expressed as*

$$\begin{aligned} \mathcal{E}_* &= \{h + e \in \mathcal{H} \oplus \mathcal{E} \colon \|Ch + De\|^2 = \|h\|^2 + \|e\|^2\} \\ &= \{h + e \in \mathcal{H} \oplus \mathcal{E} \colon A_j h + B_j e = 0 \textit{ for } j = 1, \dots, d\}. \end{aligned} \tag{3.3.10}$$

PROOF. From (3.1.9) we see that $\mathcal{E}_*$ can be expressed as

$$\mathcal{E}_* = [\mathcal{H} \oplus \mathcal{E}] \ominus \begin{bmatrix} \mathcal{U}_1 & \dots & \mathcal{U}_d \end{bmatrix} (\oplus_{j=1}^d \mathcal{H}). \tag{3.3.11}$$

Given $h + e \in \mathcal{H} \oplus \mathcal{E}$, we therefore have $h + e \in \mathcal{E}_*$ if and only if

$$\|Ch + De\|^2 = \|P_{\mathcal{E}_*}(h+e)\|^2 = \|h+e\|^2 = \|h\|^2 + \|e\|^2.$$

From the unitary property of U, we see that this last condition in turn is equivalent to $A_j h + B_j e = 0$ for each $j = 1, \dots, d$. Alternatively, from (3.3.11) we see that $h + e \in \mathcal{H} \oplus \mathcal{E}$ is in $\mathcal{E}_*$ if and only if

$$\begin{bmatrix} \mathcal{U}_1^* \\ \vdots \\ \mathcal{U}_d^* \end{bmatrix} (h+e) \perp \oplus_{j=1}^d \mathcal{H} \Longleftrightarrow A_j h + B_j e = P_{\mathcal{H}} \mathcal{U}_j^* (h+e) = 0 \text{ for } j = 1, \dots, d.$$

In any case the two alternate formulas (3.3.10) for $\mathcal{E}_*$ follow. □

The next structural issue which we address is how minimality of the scattering system $\mathfrak{S}$ is reflected in the unitary colligation $U(\mathfrak{S})$. For this purpose, given a unitary colligation (3.3.1), define a subspace $\mathcal{H}_0$ by

$$\mathcal{H}_0 = \text{the smallest subspace of } \mathcal{H} \text{ invariant for each of } A_1, \dots, A_d, A_1^*, \dots, A_d^*$$
$$\text{and containing } \operatorname{im} B_1, \dots, \operatorname{im} B_d, \operatorname{im} C^*. \tag{3.3.12}$$

We say that U is *closely connected* if the subspace $\mathcal{H}_0$ turns out to be the whole space $\mathcal{H}$. Then we have the following result.

THEOREM 3.3.3. *If $\mathfrak{S}$ is a Cuntz scattering system with associated unitary colligation $U(\mathfrak{S})$ as in Theorem 3.3.1, then $\mathfrak{S}$ is minimal if and only if $U(\mathfrak{S})$ is closely connected.*

PROOF. The result will follow from the following more general result.

THEOREM 3.3.4. *Let $\mathfrak{S}$ be a Cuntz scattering system with associated unitary colligation $U(\mathfrak{S})$. Define the following subspaces of the scattering space $\mathcal{H}$ for $\mathfrak{S}$ (equal to the state space for $U(\mathfrak{S})$):*

$$\mathcal{H}_{u1} = \text{the largest subspace of } \mathcal{H} \text{ invariant for each of } A_1, \dots, A_d, A_1^*, \dots, A_d^*$$
$$\text{for which the operator } d\text{-tuple } (A_1^*|_{\mathcal{H}_{u1}}, \dots, A_d^*|_{\mathcal{H}_{u1}}) \text{ is row-unitary,}$$
$$\mathcal{H}_{u2} = \mathcal{H}_0^\perp,$$
$$\mathcal{H}_{u3} = [\widetilde{\mathcal{G}}_* + \widetilde{\mathcal{G}}]^\perp.$$

Then $\mathcal{H}_{u1} = \mathcal{H}_{u2} = \mathcal{H}_{u3}$.

PROOF. We show that $\mathcal{H}_{u1} \subset \mathcal{H}_{u2} \subset \mathcal{H}_{u3} \subset \mathcal{H}_{u1}$.

To show that $\mathcal{H}_{u1} \subset \mathcal{H}_{u2}$ it suffices to show that $\mathcal{H}_{u1} \perp \operatorname{im} B_j$ for each $j = 1, \dots, d$ and $\mathcal{H}_{u1} \perp \operatorname{im} C^*$. For $h \in \mathcal{H}_{u1}$, the row-unitary property of the operator $(A_1^*|_{\mathcal{H}_{u1}}, \dots, A_d^*|_{\mathcal{H}_{u1}})$ implies that

$$(A_1^* A_1 + \cdots + A_d^* A_d) h = h.$$

On the other hand, from the unitary property of U we have

$$(A_1^* A_1 + \cdots + A_d^* A_d + C^* C) h = h$$

from which we get $Ch = 0$, or $\mathcal{H}_{u1} \subset \ker C$. Hence $\mathcal{H}_{u1} \perp \operatorname{im} C^*$ as wanted. Similarly, for each $j = 1, \dots, d$, the the row-unitary property of $(A_1^*|_{\mathcal{H}_{u1}}, \dots, A_d^*|_{\mathcal{H}_{u1}})$ implies that

$$A_j A_j^* h = h$$

while the unitary property of U implies

$$(A_j A_j^* + B_j B_j^*) h = h$$

from which we get $\mathcal{H}_{u1} \subset \ker B_j^*$, or $\mathcal{H}_{u1} \perp \operatorname{im} B_j$, and the inclusion $\mathcal{H}_{u1} \subset \mathcal{H}_{u2}$ follows.

To verify that $\mathcal{H}_{u2} \subset \mathcal{H}_{u3}$, we actually show that $\mathcal{H}_{u3}^{\perp} \subset \mathcal{H}_{u2}^{\perp}$. Note that $\mathcal{H}_{u3}^{\perp}$ and $\mathcal{H}_{u2}^{\perp}$ are given by

$$\begin{aligned} \mathcal{H}_{u3}^{\perp} &= \text{cl. span}\, P_{\mathcal{H}}\{\mathcal{U}^{\alpha}\mathcal{U}^{*\beta}e,\ \mathcal{U}^{\alpha}\mathcal{U}^{*\beta}e_*\colon \alpha,\beta \in \mathcal{F}_d, e \in \mathcal{E}, e_* \in \mathcal{E}_*\}, \\ \mathcal{H}_{u2}^{\perp} &= \text{cl. span}\{A^{*\alpha}A^{\beta}B_j e, A^{*\alpha}C^*e_*\colon \alpha,\beta \in \mathcal{F}_d, e \in \mathcal{E}, e_* \in \mathcal{E}_*, j = 1,\dots,d\}. \end{aligned} \tag{3.3.13}$$

In the following discussions, let $j \in \{1,\dots,d\}$, $\alpha,\beta \in \mathcal{F}_d$, $e \in \mathcal{E}$ and $e_* \in \mathcal{E}_*$. Note first that $P_{\mathcal{H}}e = 0 \in \mathcal{H}_{u2}^{\perp}$. Then note $P_{\mathcal{H}}\mathcal{U}_j^*e = B_j e \in \mathcal{H}_{u2}^{\perp}$. From (3.1.10) we see that

$$\mathcal{U}_j^* e \in \mathcal{U}_j^*\mathcal{E}_* \oplus \mathcal{H}$$

where

$$\mathcal{U}_j^*\mathcal{E}_* \subset \mathcal{U}_j^* \begin{bmatrix} \mathcal{U}_1 & \dots & \mathcal{U}_d \end{bmatrix} \left[\oplus_{j=1}^d \mathcal{G}_*\right] = \mathcal{G}_* \tag{3.3.14}$$

Combine this with the invariance of $\mathcal{G}_*$ under each of $\mathcal{U}_1^*,\dots,\mathcal{U}_d^*$ to get

$$P_{\mathcal{H}}\mathcal{U}^{*\beta g_j} = P_{\mathcal{H}}\mathcal{U}^{*\beta}\mathcal{U}_j^*e = P_{\mathcal{H}}\mathcal{U}^{*\beta}P_{\mathcal{H}}\mathcal{U}_j^*e = A^{\beta}B_j e \in \mathcal{H}_0'.$$

Similarly, note that

$$P_{\mathcal{H}}e_* = C^*e_* \in \operatorname{im} C^* \subset \mathcal{H}_{u2}^{\perp}.$$

For $\beta \neq \emptyset$, $\mathcal{U}^{*\beta}e_* \in \mathcal{G}_*$ (as noted in (3.3.14)), and hence

$$P_{\mathcal{H}}\mathcal{U}^{*\beta}e_* = 0 \in \mathcal{H}_{u2}^{\perp}.$$

Inductively, suppose that

$$P_{\mathcal{H}}\mathcal{U}^{\alpha}\mathcal{U}^{*\beta}e_* \in \mathcal{H}_{u2}^{\perp}$$

for all $\alpha,\beta \in \mathcal{F}_d$ with $|\alpha| < N$. Then for any α,β with $|\alpha| < N$ we have

$$\begin{aligned} P_{\mathcal{H}}\mathcal{U}_j\mathcal{U}^{\alpha}\mathcal{U}^{*\beta}e_* &= P_{\mathcal{H}}\mathcal{U}_j(P_{\mathcal{G}_*} + P_{\mathcal{H}})\mathcal{U}^{\alpha}\mathcal{U}^{*\beta}e_* \\ &= P_{\mathcal{H}}P_{[\mathcal{U}_1\ \dots\ \mathcal{U}_d][\oplus_{j=1}^d \mathcal{G}_*]}\mathcal{U}_jP_{\mathcal{G}_*}\mathcal{U}^{\alpha}\mathcal{U}^{*\beta}e_* + A_jP_{\mathcal{H}}\mathcal{U}^{\alpha}\mathcal{U}^{*\beta}e_* \\ &= P_{\mathcal{H}}P_{\mathcal{E}_*}\mathcal{U}_jP_{\mathcal{G}_*}\mathcal{U}^{\alpha}\mathcal{U}^{*\beta}e_* + A_jP_{\mathcal{H}}\mathcal{U}^{\alpha}\mathcal{U}^{*\beta}e_* \\ &= C^*P_{\mathcal{E}_*}\mathcal{U}_jP_{\mathcal{G}_*}\mathcal{U}^{\alpha}\mathcal{U}^{*\beta}e_* + A_jP_{\mathcal{H}}\mathcal{U}^{\alpha}\mathcal{U}^{*\beta}e_*. \end{aligned} \tag{3.3.15}$$

From the identities

$$A_jA_k^* = \delta_{j,k}I - B_jB_k^*, \qquad A_jC^* = -B_jD^*$$

arising from the unitary property of U, we see that $A_jP_{\mathcal{H}}\mathcal{U}^{\alpha}\mathcal{U}^{*\beta}e_* \in \mathcal{H}_{u2}^{\perp}$ whenever $P_{\mathcal{H}}\mathcal{U}^{\alpha}\mathcal{U}^{*\beta}e_* \in \mathcal{H}_{u2}^{\perp}$. Trivially we have $\operatorname{im} C^* \in \mathcal{H}_{u2}^{\perp}$ as well, and hence it follows from (3.3.15) that $P_{\mathcal{H}}\mathcal{U}^{\alpha}\mathcal{U}^{*\beta}e_* \in \mathcal{H}_{u2}^{\perp}$ for all α and β, and it follows that $\mathcal{H}_{u2} \subset \mathcal{H}_{u3}$.

We next show that $\mathcal{H}_{u3} \subset \mathcal{H}_{u1}$. Since $\widetilde{\mathcal{G}}_* + \widetilde{\mathcal{G}}$ is reducing for each $\mathcal{U}_j$, it follows that $\mathcal{H}_{u3}$ is reducing for each $\mathcal{U}_j$. Moreover, $\mathcal{H}_{u3} \subset \mathcal{H}$. Hence the projection is not needed and we have $\mathcal{U}_j|_{\mathcal{H}_{u3}} = A_j^*|_{\mathcal{H}_{u3}}$ and $\mathcal{U}_j^*|_{\mathcal{H}_{u3}} = A_j|_{\mathcal{H}_{u3}}$ for each j. It follows therefore that $(A_1^*|_{\mathcal{H}_{u3}},\dots,A_d^*|_{\mathcal{H}_{u3}})$ is row-unitary. Then by definition of $\mathcal{H}_{u1}$, we see that $\mathcal{H}_{u3} \subset \mathcal{H}_{u1}$, and Theorem 3.3.4 follows. □

To complete the proof of Theorem 3.3.3, note that $\mathfrak{S}$ is minimal if and only if $\mathcal{H}_{u3} = \{0\}$, and that $U(\mathfrak{S})$ is closely connected if and only if $\mathcal{H}_{u2} = \{0\}$. By Theorem 3.3.4 it is always the case that $\mathcal{H}_{u3} = \mathcal{H}_{u2}$, and hence the equivalence of minimality of $\mathfrak{S}$ with close-connectedness of $U(\mathfrak{S})$ follows. □

It is of some interest that the unitary subspace $\mathcal{H}_u$ (equal to the common value of $\mathcal{H}_{u1}$, $\mathcal{H}_{u2}$ and $\mathcal{H}_{u3}$ in Theorem 3.3.4) for the row-contraction $(A_1^*, \dots, A_d^*)$ has the following more explicit description.

PROPOSITION 3.3.5. *Let U be a unitary colligation as in (3.3.1) and let $\mathcal{H}_u$ be the subspace $\mathcal{H}_{u1} = \mathcal{H}_{u2} = \mathcal{H}_{u3}$ in Theorem 3.3.4. Then $\mathcal{H}_u$ can also be described as*

$$\begin{aligned}\mathcal{H}_u =&\{h \in \mathcal{H}\colon \|A^v h\|^2 = \sum_{k=1}^d \|A_k A^v h\|^2 = \|A^{*w} A^v h\|^2 \\ &\text{for all } v, w \in \mathcal{F}_d \text{ and } N = 0, 1, 2, \dots\},\end{aligned} \tag{3.3.16}$$

or equivalently as

$$\mathcal{H}_u = \left[\bigcap_{\alpha,\beta,j} \ker B_j^* A^{*\beta} A^\alpha\right] \bigcap \left[\bigcap_\alpha \ker C A^\alpha\right]. \tag{3.3.17}$$

PROOF. Temporarily set the right-hand side of (3.3.16) equal to $\mathcal{H}'_u$ and the right-hand side of (3.3.17) equal to $\mathcal{H}''_u$. From the definitions and a simple duality argument, we see that $\mathcal{H}''_u = \mathcal{H}_{u2}$. Since $(A_1^*|_{\mathcal{H}_u}, \dots, A_d^*|_{\mathcal{H}_{u1}})$ is row unitary, it follows that $\mathcal{H}_{u1} \subset \mathcal{H}'_u$, and hence $\mathcal{H}''_u = \mathcal{H}_{u2} = \mathcal{H}_{u1} \subset \mathcal{H}'_u$. It remains therefore only to show that $\mathcal{H}'_u \subset \mathcal{H}''_u$.

Recalling that

$$U = \begin{bmatrix} A_1 & B_1 \\ \vdots & \vdots \\ A_d & B_d \\ C & D \end{bmatrix}$$

is unitary, for $h \in \mathcal{H}'_u$ we compute

$$\begin{aligned}\|B_j A^{*\beta} A^\alpha h\|^2 &= \langle A^{*\alpha^\top} A^{\beta^\top} B_j B_j^* A^{*\beta} A^\alpha h, h\rangle \\ &= A^{*\alpha^\top} A^{\beta^\top} (I - A_j A_j^*) A^{*\beta} A^\alpha h, h\rangle \\ &= \|A^{*\beta} A^\alpha h\|^2 - \|A^{*g_j\beta} A^\alpha h\|^2 \\ &= \|A^\alpha h\|^2 - \|A^\alpha h\|^2 \text{ (since } h \in \mathcal{H}'_u) \\ &= 0.\end{aligned}$$

Similarly,

$$\begin{aligned}\|CA^\alpha h\|^2 &= \langle A^{*\alpha^\top} C^* C A^\alpha h, h\rangle \\ &= \langle A^{*\alpha^\top} (I - A_1^* A_1 - \cdots - A_d^* A_d) A^\alpha h, h\rangle \\ &= \|A^\alpha h\|^2 - \sum_{k=1}^d \|A_k A^\alpha h\|^2 \\ &= 0 \text{ (since } h \in \mathcal{H}'_u)\end{aligned}$$

and it follows that $h \in \mathcal{H}''_u$ as wanted. □

REMARK 3.3.6. We note that the decomposition $[\mathcal{H} \oplus \mathcal{G}] = [\mathcal{H}_0 \oplus \mathcal{G}] \oplus \mathcal{H}_u$ is just the Wold decomposition for the row isometry $(\mathcal{U}_1|_{\mathcal{H}\oplus\mathcal{G}}, \dots, \mathcal{U}_d|_{\mathcal{H}\oplus\mathcal{G}})$ (see [**Po89c**]).

In Section 3.1 we introduced the Fourier representation operators $\Phi\colon \mathcal{K} \to \mathcal{L}_W$, $\Phi_*\colon \mathcal{K} \to \mathcal{L}_{W_*}$ and $\Phi_{\mathcal{H}}\colon \mathcal{K} \to \mathcal{L}_{W_{\mathcal{H}}}$ (see (3.1.17)). We now work with "time-domain" versions (see Section 4.1 below) of Φ, Φ_* and $\Phi_{\mathcal{H}}$. Specifically, for $v, w \in \mathcal{F}_d$ introduce operators

$$\begin{aligned} \Phi_{v,w} &= P_{\mathcal{E}}\mathcal{U}^w\mathcal{U}^{*v}\colon \mathcal{K} \to \mathcal{E} \\ \Phi_{*v,w} &= P_{\mathcal{E}_*}\mathcal{U}^w\mathcal{U}^{*v}\colon \mathcal{K} \to \mathcal{E}_*, \\ \Phi_{\mathcal{H},v,w} &= P_{\mathcal{H}}\mathcal{U}^w\mathcal{U}^{*v}\colon \mathcal{K} \to \mathcal{H}. \end{aligned}$$

Note that these operators are the maps defining the Fourier coefficients for the operators Φ, Φ_* and $\Phi_{\mathcal{H}}$:

$$\begin{aligned} \Phi\colon k &\mapsto \sum_{v,w\in\mathcal{F}_d} (\Phi_{v,w}k) z^v \zeta^w, \\ \Phi_*\colon k &\mapsto \sum_{v,w\in\mathcal{F}_d} (\Phi_{*v,w}k) z^v \zeta^w, \\ \Phi_{\mathcal{H}}\colon k &\mapsto \sum_{v,w\in\mathcal{F}_d} (\Phi_{\mathcal{H},v,w}k) z^v \zeta^w. \end{aligned}$$

We define operators $\widetilde{\Phi}\colon \mathcal{K} \to \ell_W \subset \ell(\mathcal{F}_d \times \mathcal{F}_d, \mathcal{E})$, $\widetilde{\Phi}_*\colon \mathcal{K} \to \ell_{W_*}$ and $\widetilde{\Phi}_{\mathcal{H}}\colon \mathcal{K} \to \ell_{W_{\mathcal{H}}}$ by

$$\widetilde{\Phi}\colon k \mapsto u, \qquad \widetilde{\Phi}_*\colon k \mapsto y, \qquad \widetilde{\Phi}_{\mathcal{H}}\colon k \mapsto x$$

where

$$\begin{aligned} u(v,w) &= (\widetilde{\Phi}k)(v,w) = P_{\mathcal{E}}\mathcal{U}^w\mathcal{U}^{*v}k, \\ y(v,w) &= (\widetilde{\Phi}_*k)(v,w) = P_{\mathcal{E}_*}\mathcal{U}^w\mathcal{U}^{*v}k, \\ x(v,w) &= (\widetilde{\Phi}_{\mathcal{H}}k)(v,w) = P_{\mathcal{H}}\mathcal{U}^w\mathcal{U}^{*v}k. \end{aligned}$$

Here, for a positive semidefinite Cuntz weight W, we define ℓ_W as the image of $\mathcal{L}_W$ under the inverse Z-transform, so $\widehat{\ell_W} = \mathcal{L}_W$; here, in general the *$Z$-transform* is the mapping from a vector-valued function $\{f_{v,w}\}_{v,w\in\mathcal{F}_d}$ defined on the "time-axis" $\mathcal{F}_d \times \mathcal{F}_d$ to the associated formal power series $\widehat{f}(z,\zeta)$ given by

$$\{f(v,w)\}_{v,w\in\mathcal{F}_d} \mapsto \widehat{f}(z,\zeta) := \sum_{v,w\in\mathcal{F}_d} f(v,w) z^v \zeta^w$$

and will be discussed further in Section 4.1 where we introduce the connections with system theory. Similarly, we write h_W so that $\widehat{h_W} = \mathcal{H}_W$, and $h_W^\perp = \ell_W \ominus h_W$ so that $(h_W^\perp)^\wedge = \mathcal{H}_W^\perp$ when needed. We denote the mapping corresponding to the whole aggregate of $\widetilde{\Phi}$, $\widetilde{\Phi}_{\mathcal{H}}$ and $\widetilde{\Phi}_*$ by Ω:

$$\Omega\colon k \mapsto (u, x, y) := (\widetilde{\Phi}k,\ \widetilde{\Phi}_{\mathcal{H}}k,\ \widetilde{\Phi}_*k). \tag{3.3.18}$$

Throughout this section we shall also need the time-domain versions $\widetilde{S}_j^R$, $\widetilde{U}_j^R$, $\widetilde{S}_j^{R[*]}$ and $\widetilde{U}_j^{R[*]}$ of the shift operators S_j^R, U_j^R, $S_j^{[*]}$ and $U_j^{R[*]}$ defined in (2.2.13), (2.2.15),

(2.2.17) and (2.2.16); these are

$$\widetilde{S}_j^R \colon \{f(v,w)\}_{v,w\in\mathcal{F}_d} \mapsto \{f(vg_j^{-1},w)\}_{v,w\in\mathcal{F}_d}, \tag{3.3.19}$$

$$\widetilde{U}_j^R \colon \{f(v,w)\}_{v,w\in\mathcal{F}_d} \mapsto \begin{cases} \{f(\emptyset, wg_j)\}_{v,w\in\mathcal{F}_d} & \text{if } v=\emptyset, \\ \{f(vg_j^{-1},w)\}_{v,w\in\mathcal{F}_d} & \text{if } v\neq\emptyset, \end{cases} \tag{3.3.20}$$

$$\widetilde{S}_j^{R[*]} \colon \{f(v,w)\}_{v,w\in\mathcal{F}_d} \mapsto \{f(vg_j,w)\}_{v,w\in\mathcal{F}_d}, \tag{3.3.21}$$

$$\widetilde{U}_j^{R[*]} \colon \{f(v,w)\}_{v,w\in\mathcal{F}_d} \mapsto \begin{cases} \{f(\emptyset, wg_j^{-1})\}_{v,w\in\mathcal{F}_d} & \text{if } v=\emptyset, \\ \{f(vg_j,w)\}_{v,w\in\mathcal{F}_d} & \text{if } v\neq\emptyset. \end{cases} \tag{3.3.22}$$

A consequence of the general intertwining relation (2.2.10) combined with the explicit formulas (2.2.20) for $\mathcal{U}_{W,j}$ and $\mathcal{U}_{W,j}^*$ is that Ω intertwines the unitary evolution operator $\mathcal{U}_j$ with the bilateral forward shift U_j^R on each component of the image of Ω and similarly for the adjoint: namely, we have

$$\Omega\colon k\mapsto (u,x,y) \Longrightarrow \begin{cases} \Omega\colon \mathcal{U}_j k \mapsto (\widetilde{U}_j^R u, \widetilde{U}_j^R x, \widetilde{U}_j^R y) \\ \Omega\colon \mathcal{U}_j^* k \mapsto (\widetilde{S}_j^{R[*]} u, \widetilde{S}_j^{R[*]} x, \widetilde{S}_j^{R[*]} y). \end{cases} \tag{3.3.23}$$

Then we have the following remarkable result.

THEOREM 3.3.7. *Let $\mathfrak{S}$ be a Cuntz scattering system with associated unitary colligation $U = U(\mathfrak{S})$ as in (3.3.2), and define the Cuntz weights W and W_* by (3.1.14) and (3.1.15). Then, for k an element of the ambient space $\mathcal{K}$ of $\mathfrak{S}$, the $\mathcal{E}\times\mathcal{H}\times\mathcal{E}_*$-valued function $\Omega(k) = (u,x,y)$ has first component u in the space ℓ_W and third component y in the space ℓ_{W_*}, and forms a trajectory of the system $\Sigma(U)$ associated with U, in the sense that (u,x,y) satisfies the systems of recursions*

$$\begin{aligned} x(g_j v,\emptyset) &= A_j x(v,\emptyset) + B_j u(v,\emptyset) \\ y(v,\emptyset) &= C x(v,\emptyset) + D u(v,\emptyset) \end{aligned} \tag{3.3.24}$$

and

$$\begin{aligned} x(v,g_j w) &= A_j^* x(v,w) + C^* y(v,g_j w) \\ u(v,g_j w) &= B_j^* x(v,w) + D^* y(v,g_j w) \end{aligned} \tag{3.3.25}$$

for $j=1,\dots,d$.

Conversely, any solution (u,x,y) of the systems of equations (3.3.24) and (3.3.25) such that $u\in\ell_W$ and $y\in\ell_{W_*}$ is necessarily of the form $(u,x,y)=\Omega(k)$ for some k in the ambient space $\mathcal{K}$ of $\mathfrak{S}$.

PROOF. From the scattering axioms and the decompositions

$$\begin{aligned} \mathcal{K} &= \mathcal{G}_* \oplus \mathcal{H} \oplus \mathcal{G} \\ &= \mathcal{G}_* \oplus \mathcal{H} \oplus \mathcal{E} \oplus \begin{bmatrix} \mathcal{U}_1 & \dots & \mathcal{U}_d \end{bmatrix} \left[\oplus_{j=1}^d \mathcal{G}\right] \\ &= \mathcal{G}_* \oplus \mathcal{E}_* \oplus \begin{bmatrix} \mathcal{U}_1 & \dots & \mathcal{U}_d \end{bmatrix} \left[\oplus_{j=1}^d \mathcal{H}\right] \oplus \begin{bmatrix} \mathcal{U}_1 & \dots & \mathcal{U}_d \end{bmatrix} \left[\oplus_{j=1}^d \mathcal{G}\right] \end{aligned}$$

(see (3.1.9)), one can easily verify the basic identities

$$P_{\mathcal{H}}\mathcal{U}_k^* = P_{\mathcal{H}}\mathcal{U}_k^*(P_{\mathcal{H}} + P_{\mathcal{E}}), \tag{3.3.26}$$

$$P_{\mathcal{E}_*} = P_{\mathcal{E}_*}(P_{\mathcal{H}} + P_{\mathcal{E}}). \tag{3.3.27}$$

These will be used repeatedly in the sequel.

Let now $k \in \mathcal{K}$ and set

$$x(v,w) = P_{\mathcal{H}}\mathcal{U}^{w}\mathcal{U}^{*v}k, \qquad u(v,w) = P_{\mathcal{E}}\mathcal{U}^{w}\mathcal{U}^{*v}k, \qquad y(v,w) = P_{\mathcal{E}_*}\mathcal{U}^{w}\mathcal{U}^{*v}k$$

From the discussion in Chapter 2 (we refer to [**BaV04**] for complete details), we know that $(\Phi k)(z,\zeta) \in \mathcal{L}_W$ and $(\Phi_* k)(z,\zeta) \in \mathcal{L}_{W_*}$; changing notation from the frequency domain to the time domain then gives that $u \in \ell_W$ and $y \in \ell_{W_*}$. Next we compute

$$\begin{aligned} x(g_j v, \emptyset) &= P_{\mathcal{H}}\mathcal{U}_j^{*}\mathcal{U}^{*v}k \\ &= P_{\mathcal{H}}\mathcal{U}_j^{*}(P_{\mathcal{H}} + P_{\mathcal{E}})\mathcal{U}^{*v}k \text{ (by (3.3.26))} \\ &= A_j P_{\mathcal{H}}\mathcal{U}^{*v}k + B_j P_{\mathcal{E}}\mathcal{U}^{*v}k \\ &= A_j x(v,\emptyset) + B_j u(v,\emptyset). \end{aligned}$$

while

$$\begin{aligned} y(v,\emptyset) &= P_{\mathcal{E}_*}\mathcal{U}^{*v}k \\ &= P_{\mathcal{E}_*}(P_{\mathcal{H}} + P_{\mathcal{E}})\mathcal{U}^{*v}k \text{ (by (3.3.27))} \\ &= C P_{\mathcal{H}}\mathcal{U}^{*v}k + D P_{\mathcal{E}}\mathcal{U}^{*v}k \\ &= Cx(v,\emptyset) + Du(v,\emptyset) \end{aligned}$$

and the system of recursions (3.3.24) follows.

On the other hand,

$$\begin{aligned} A_j x(v,w) + B_j u(v,w) &= P_{\mathcal{H}}\mathcal{U}_j^{*} P_{\mathcal{H}}\mathcal{U}^{w}\mathcal{U}^{*v}k + P_{\mathcal{H}}\mathcal{U}_j^{*} P_{\mathcal{E}}\mathcal{U}^{w}\mathcal{U}^{*v}k \\ &= P_{\mathcal{H}}\mathcal{U}_j^{*}(P_{\mathcal{H}} + P_{\mathcal{E}})\mathcal{U}^{w}\mathcal{U}^{*v}k \\ &= P_{\mathcal{H}}\mathcal{U}_j^{*}\mathcal{U}^{w}\mathcal{U}^{*v}k \text{ (by (3.3.26))} \\ &= \begin{cases} 0 & \text{if } w \text{ does not end with } g_j, \\ x(v, g_j^{-1}w) & \text{otherwise,} \end{cases} \end{aligned} \tag{3.3.28}$$

while

$$\begin{aligned} Cx(v,w) + Du(v,w) &= P_{\mathcal{E}_*} P_{\mathcal{H}}\mathcal{U}^{w}\mathcal{U}^{*v}k + P_{\mathcal{E}_*} P_{\mathcal{E}}\mathcal{U}^{w}\mathcal{U}^{*v}k \\ &= P_{\mathcal{E}_*}\mathcal{U}^{w}\mathcal{U}^{*v}k \text{ (by (3.3.27))} \\ &= y(v,w). \end{aligned} \tag{3.3.29}$$

Writing (3.3.28) and (3.3.29) in matrix form gives

$$\begin{bmatrix} A_1 & B_1 \\ \vdots & \vdots \\ A_d & B_d \\ C & D \end{bmatrix} \begin{bmatrix} x(v, g_j w) \\ u(v, g_j w) \end{bmatrix} = \begin{bmatrix} 0 \\ \vdots \\ 0 \\ x(v,w) \\ 0 \\ \vdots \\ 0 \\ y(v, g_j w) \end{bmatrix} \tag{3.3.30}$$

where the first nonzero entry in the column on the right hand side occurs in the jth slot. As the matrix

$$U = \begin{bmatrix} A_1 & B_1 \\ \vdots & \vdots \\ A_d & B_d \\ C & D \end{bmatrix}$$

is unitary, we can multiply both sides of (3.3.30) by U^* to get

$$\begin{bmatrix} A_1^* & \dots & A_d^* & C^* \\ B_1^* & \dots & B_d^* & D^* \end{bmatrix} \begin{bmatrix} 0 \\ \vdots \\ 0 \\ x(v,w) \\ 0 \\ \vdots \\ 0 \\ y(v, g_j w) \end{bmatrix} = \begin{bmatrix} x(v, g_j w) \\ u(v, g_j w) \end{bmatrix}.$$

and (3.3.25) follows.

Now suppose that (u, x, y) is any solution of (3.3.24) and (3.3.25) such that $u \in \ell_W$ and $y \in \ell_{W_*}$. Define $k \in \mathcal{K}$ by

$$k = \widetilde{\Phi}_*^*(P_{h_{W_*}^\perp} y) + x(\emptyset, \emptyset) + \widetilde{\Phi}^*(P_{h_W} u).$$

Then $k \in \mathcal{K} = \mathcal{G}_* \oplus \mathcal{H} \oplus \mathcal{G}$ with $\mathcal{G}_*$-component equal to $k_{\mathcal{G}_*} := \widetilde{\Phi}_*^*(P_{h_{W_*}^\perp} y)$, with $\mathcal{H}$-component equal to $k_{\mathcal{H}} := x(\emptyset, \emptyset)$ and with $\mathcal{G}$-component equal to $k_{\mathcal{G}} := \widetilde{\Phi}^* P_{h_W} u$. Set $(u', x', y') = \Omega(k)$, i.e.,

$$\begin{aligned} u'(v,w) &= (\widetilde{\Phi} k)(v,w) = P_{\mathcal{E}} \mathcal{U}^w \mathcal{U}^{*v} k, \\ y'(v,w) &= (\widetilde{\Phi}_* k)(v,w) = P_{\mathcal{E}_*} \mathcal{U}^w \mathcal{U}^{*v} k, \\ x'(v,w) &= (\widetilde{\Phi}_{\mathcal{H}} k)(v,w) = P_{\mathcal{H}} \mathcal{U}^w \mathcal{U}^{*v} k. \end{aligned}$$

The problem is to show that $(u', x', y') = (u, x, y)$. It is immediately obvious that

$$x'(\emptyset, \emptyset) = x(\emptyset, \emptyset). \tag{3.3.31}$$

Next observe that, for $v \in \mathcal{F}_d$,

$$\begin{aligned} u'(v, \emptyset) &= P_{\mathcal{E}} \mathcal{U}^{*v} k \\ &= P_{\mathcal{E}} \mathcal{U}^{*v} k_{\mathcal{G}_*} + P_{\mathcal{E}} \mathcal{U}^{*v} k_{\mathcal{H}} + P_{\mathcal{E}} \mathcal{U}^{*v} k_{\mathcal{G}} \\ &= P_{\mathcal{E}} \mathcal{U}^{*v} \widetilde{\Phi}^* P_{h_W} u \\ &= (\widetilde{\Phi} \widetilde{\Phi}^* P_{h_W} u)(v, \emptyset) \\ &= (P_{h_W} u)(v, \emptyset) \text{ (since } \widetilde{\Phi} \colon \mathcal{K} \to \ell_W \text{ is a coisometry)} \\ &= u(v, \emptyset) \text{ (by (2.3.8))}. \end{aligned}$$

and hence

$$u'(v, \emptyset) = u(v, \emptyset) \text{ for all } v \in \mathcal{F}_d. \tag{3.3.32}$$

By the first part of the proof, we know that (u', x', y') solves (3.3.24). Then from (3.3.31) and (3.3.32) we see that

$$x'(v, \emptyset) = x(v, \emptyset) \text{ and } y'(v, \emptyset) = y(v, \emptyset) \text{ for all } v \in \mathcal{F}_d. \tag{3.3.33}$$

As a consequence of the identity (2.3.8) we then get from (3.3.33) that

$$P_{h_{W_*}} y' = P_{h_{W_*}} y. \tag{3.3.34}$$

Next we observe, for all $v, w \in \mathcal{F}_d$,

$$\begin{aligned}(P_{h^\perp_{W_*}} y')(v, w) &= (\widetilde{\Phi}_* P_{\mathcal{G}_*} k)(v, w) \\ &= (\widetilde{\Phi}_* \widetilde{\Phi}_*^* P_{h^\perp_{W_*}} y)(v, w) \\ &= (P_{h^\perp_{W_*}} y)(v, w) \text{ (since } \widetilde{\Phi}_* \colon \mathcal{K} \to \ell_{W_*} \text{ is a coisometry)}\end{aligned}$$

and hence

$$P_{h^\perp_{W_*}} y' = P_{h^\perp_{W_*}} y. \tag{3.3.35}$$

Combining (3.3.34) and (3.3.35) gives

$$y'(v, w) = y(v, w) \text{ for all } v, w \in \mathcal{F}_d. \tag{3.3.36}$$

By the first part of the proof, we know that (u', x', y') satisfies (3.3.25). As (u, x, y) also satisfies (3.3.25) by hypothesis, from (3.3.31) combined with (3.3.36) (used for the special case where $w \neq \emptyset$) we are forced to conclude that

$$x'(v, w) = x(v, w) \text{ and } u'(v, w) = u(v, w) \text{ for all } v, w \in \mathcal{F}_d \text{ with } w \neq \emptyset. \tag{3.3.37}$$

Combining (3.3.32), (3.3.33), (3.3.36) and (3.3.37) finally leads us to the conclusion $(u', x', y') = (u, x, y)$ as wanted. This completes the proof of Theorem 3.3.7. □

We have the following characterization of the norm of an element k of the ambient space $\mathcal{K}$ in terms of the associated system trajectory $(u, x, y) = \Omega k$.

THEOREM 3.3.8. *Let k be an element of the ambient space $\mathcal{K}$ of a Cuntz scattering system $\mathfrak{S}$ with associated system trajectory $(u, x, y) = \Omega k$ as in Proposition 3.3.7. Then*

$$\begin{aligned}\|k\|^2 &= \|P_{h^\perp_{W_*}} y\|^2_{h^\perp_{W_*}} + \|x(\emptyset, \emptyset)\|^2_{\mathcal{H}} + \|P_{h_W} u\|^2 \\ &= \left[\|y\|^2_{\ell_{W_*}} - \sum_{v \in \mathcal{F}_d} \|y(v, \emptyset)\|^2_{\mathcal{E}_*}\right] + \|x(\emptyset, \emptyset)\|^2 + \sum_{v \in \mathcal{F}_d} \|u(v, \emptyset)\|^2_{\mathcal{E}}.\end{aligned} \tag{3.3.38}$$

PROOF. Since $\mathcal{K} = \mathcal{G}_* \oplus \mathcal{H} \oplus \mathcal{G}$, we may write

$$\|k\|^2_{\mathcal{K}} = \|P_{\mathcal{G}_*} k\|^2 + \|P_{\mathcal{H}} k\|^2 + \|P_{\mathcal{G}} k\|^2.$$

Next note that $\widetilde{\Phi}_*$ is isometric on $\mathcal{G}_*$, that $P_{\mathcal{H}} k = x(\emptyset, \emptyset)$ by definition, that $\widetilde{\Phi}$ is isometric on $\mathcal{G}$ along with the intertwining conditions

$$\widetilde{\Phi}_* P_{\mathcal{G}_*} = P_{h^\perp_{W_*}} \widetilde{\Phi}_*, \qquad \widetilde{\Phi} P_{\mathcal{G}} = P_{h_W} \widetilde{\Phi}$$

to conclude that

$$\begin{aligned}\|P_{\mathcal{G}_*} k\|^2 &= \|P_{h^\perp_{W_*}} y\|^2, \\ \|P_{\mathcal{H}} k\|^2 &= \|x(\emptyset, \emptyset)\|^2, \\ \|P_{\mathcal{G}} k\|^2 &= \|P_{h_W} u\|^2.\end{aligned}$$

Finally from the identity (2.3.10) we know that

$$\begin{aligned}\|P_{h_{W_*}^{\perp}} y\|^2 &= \|y\|^2_{\ell_{W_*}} - \|P_{h_{W_*}} y\|^2 \\ &= \|y\|^2_{\ell_{W_*}} - \sum_{v \in \mathcal{F}_d} \|y(v, \emptyset)\|^2_{\mathcal{E}_*}, \\ \|P_{h_W} u\|^2 &= \sum_{v \in \mathcal{F}_d} \|u(v, \emptyset)\|^2_{\mathcal{E}}.\end{aligned}$$

This concludes the proof of Theorem 3.3.8. □

From the formula (3.3.38) for the norm of $k \in \mathcal{K}$, we see that a trajectory $\Omega(k) = (u, x, y)$ in $\Omega\mathcal{K}$ is uniquely determined from an element $y_{\perp} = P_{h_{W_*}^{\perp}} y$ in $h_{W_*}^{\perp}$, an element $h = x(\emptyset, \emptyset)$ of $\mathcal{H}$ together with an element $u_+ = u|_{\mathcal{F}_d \times \{\emptyset\}}$ of $\ell^2(\mathcal{F}_d, \mathcal{E})$. Thus it appears that the pair (U, W_*) is enough to determine $\mathfrak{S}$—in particular, the outgoing Cuntz weight W_* and the scattering function S appear to be completely determined from the Cuntz unitary colligation $U = U(\mathfrak{S})$ and the incoming Cuntz weight W_*. This will be made explicit in the next Chapter.

CHAPTER 4

Unitary Colligations

4.1. Preliminaries

By a d-variable colligation U (for our purposes here, not to be confused with the class studied in [**BaT98**]) we mean a Hilbert-space operator of the form

$$(4.1.1)\qquad U = \begin{bmatrix} A & B \\ C & D \end{bmatrix} = \begin{bmatrix} A_1 & B_1 \\ \vdots & \vdots \\ A_d & B_d \\ C & D \end{bmatrix} : \begin{bmatrix} \mathcal{H} \\ \mathcal{E} \end{bmatrix} \to \begin{bmatrix} \oplus_{j=1}^d \mathcal{H} \\ \mathcal{E}_* \end{bmatrix}$$

where $\mathcal{H}$ (the *state space*), $\mathcal{E}$ (the *input space*) and $\mathcal{E}_*$ (the *output space*) are Hilbert spaces. We will be particularly interested in the case where U has some metric properties (such as contractive, isometric, coisometric or contractive), but for the moment we discuss various properties which are independent of metric structure. In particular, when U in (4.1.1) is unitary, we shall say that U is a *d-variable Cuntz-unitary colligation.*

As has already come up in Section 3.3, we associate with U the "forward-time" system equations

$$(4.1.2)\qquad \Sigma(U)\text{ (forward)} \begin{cases} x(g_j v, \emptyset) &= A_j x(v,\emptyset) + B_j u(v,\emptyset) \text{ for } j = 1,\dots,d, \\ y(v,\emptyset) &= Cx(v,\emptyset) + Du(v,\emptyset) \end{cases}$$

and the "backward-time" system equations
(4.1.3)

$$\Sigma(U)\text{ (backward)} \begin{cases} x(v, g_j w) &= A_j^* x(v,w) + C^* y(v, g_j w) \\ u(v, g_j w) &= B_j^* x(v,w) + D^* y(v, g_j w) \text{ for } j = 1,\dots,d. \end{cases}$$

Given a value for the initial state $x(\emptyset,\emptyset)$ and the input sequence u on $\mathcal{F}_d \times \{\emptyset\}$, we can use (4.1.2) to generate $x(v,\emptyset)$ and $y(v,\emptyset)$ for all $v \in \mathcal{F}_d$. Similarly, from a knowledge of the initial state $x(\emptyset, v)$ (for each $v \in \mathcal{F}_d$) and a specification of the output string y on the past $\mathcal{F}_d \times (\mathcal{F}_d \setminus \{\emptyset\})$, one can use (4.1.3) to generate uniquely in a well-defined way values for the state $x(v,w)$ and the input $u(v,w)$ on the past, i.e., for all pairs of words v, w with $w \neq \emptyset$. By combining these, we can generate a trajectory (u, x, y) defined on all of $\mathcal{F}_d \times \mathcal{F}_d$ from a specification of u on $\mathcal{F}_d \times \{\emptyset\}$, of $x(\emptyset,\emptyset)$ and of y on $\mathcal{F}_d \times (\mathcal{F}_d \setminus \{\emptyset\})$. Note that both u and x on the past $\mathcal{F}_d \times (\mathcal{F}_d \setminus \{\emptyset\})$ are influenced by u on the future $\mathcal{F}_d \times \{\emptyset\}$ through their dependence on the initial conditions $x(v,\emptyset)$ with $v \neq \emptyset$; this phenomenon diverges from the intuition from the classical case and is quantified in the formulas in Theorem 4.1.1 below.

The "time" axis for a system of the type (4.1.2)-(4.1.3) is the set $\mathcal{F}_d \times \mathcal{F}_d$ of pairs of words (v, w). Just as for the classical discrete-time case, these systems are amenable to both "frequency-domain" and "time-domain" techniques. An input signal consists of an $\mathcal{E}$-valued function $\{u(v,w)\}_{v,w\in\mathcal{F}_d}$ defined on the time axis $\mathcal{F}_d \times \mathcal{F}_d$. Similarly, a state trajectory is an $\mathcal{H}$-valued function $\{x(v,w)\}_{v,w\in\mathcal{F}_d}$ and an output signal is an $\mathcal{E}_*$-valued function $\{y(v,w)\}_{v,w\in\mathcal{F}_d}$ defined on $\mathcal{F}_d\times\mathcal{F}_d$. For $\mathcal{E}$ any Hilbert space, let us denote the space of all $\mathcal{E}$-valued functions $\{y(v,w)\}_{v,w\in\mathcal{F}_d}$ on $\mathcal{F}_d \times \mathcal{F}_d$ by $\ell(\mathcal{F}_d \times \mathcal{F}_d, \mathcal{E})$. We denote by $\ell^2(\mathcal{F}_d \times \mathcal{F}_d, \mathcal{E})$ the set of those $u \in \ell(\mathcal{F}_d \times \mathcal{F}_d, \mathcal{E})$ which are norm square-summable: $\sum_{v,w\in\mathcal{F}_d} \|u(v,w)\|^2 < \infty$. These spaces form the environments for time-domain analysis. The domain for frequency-domain analysis, on the other hand, consists of spaces of the type $L(\mathcal{F}_d\times\mathcal{F}_d,\mathcal{E})$ and $L^2(\mathcal{F}_d \times \mathcal{F}_d, \mathcal{E})$ already introduced in Chapter 2, namely, spaces of formal power series (with norm square-integrable coefficients in the L^2-case)

$$f(z,\zeta) = \sum_{v,w\in\mathcal{F}_d} f_{v,w} z^v \zeta^w$$

with coefficients in $\mathcal{E}$. The formal Z-transform maps an element $f \in \ell(\mathcal{F}_d \times \mathcal{F}_d, \mathcal{E})$ into $\widehat{f}(z,\zeta) \in L(\mathcal{F}_d \times \mathcal{F}_d, \mathcal{E})$ according to the rule

$$\{f(v,w)\}_{v,w\in\mathcal{F}_d} \mapsto \widehat{f}(z,\zeta) := \sum_{v,w\in\mathcal{F}_d} f(v,w) z^v \zeta^w. \tag{4.1.4}$$

In this setting the Z-transform really amounts to merely a change of notation, since the infinite series is purely formal and hence the same as the list of its coefficients; however, multiplication operators defined in the frequency domain are more intuitive and suggestive than the corresponding noncommutative convolution operators defined in the time domain, as we have seen in Chapter 2.

Given a colligation U we define the space of trajectories $\mathcal{T}$ of the system $\Sigma(U)$ to consist of all $(\mathcal{E} \times \mathcal{H} \times \mathcal{E}_*)$-valued functions on the time axis $\mathcal{F}_d \times \mathcal{F}_d$

$$(v,w) \in \mathcal{F}_d \times \mathcal{F}_d \mapsto (u(v,w), x(v,w), y(v,w)) \in \mathcal{E} \times \mathcal{H} \times \mathcal{E}_*$$

which solve the system equations (4.1.2) and (4.1.3). From the recursive nature of these equations, we see that the map $\mathcal{I}'\colon \mathcal{T} \to \ell(\mathcal{F}_d, \mathcal{E}) \times \mathcal{H} \times \ell(\mathcal{F}_d \times (\mathcal{F}_d \setminus \{\emptyset\}))$ given by

$$\mathcal{I}'\colon (u,x,y) \mapsto (u|_{\mathcal{F}_d\times\{\emptyset\}}, x(\emptyset,\emptyset), y|_{\mathcal{F}_d\times(\mathcal{F}_d\setminus\{\emptyset\})}). \tag{4.1.5}$$

defines a bijective correspondence between $\mathcal{T}$ and the set of corresponding free parameters $\ell(\mathcal{F}_d, \mathcal{E}) \times \mathcal{H} \times \ell(\mathcal{F}_d \times (\mathcal{F}_d \setminus \{\emptyset\}))$. The next theorem makes this correspondence explicit. For this theorem we need the *forward Z-transform* $f^{\wedge+}$ and *backward Z-transform* of a given $f \in \ell(\mathcal{F}_d \times \mathcal{F}_d, \mathcal{E})$:

$$f^{\wedge+}(z) := \sum_{v\in\mathcal{F}_d} f(v,\emptyset) z^v, \tag{4.1.6}$$

$$f^{\wedge-}(z,\zeta) := \sum_{(v,w)\in\mathcal{F}_d\times(\mathcal{F}_d\setminus\{\emptyset\})} f(v,w) z^v \zeta^w. \tag{4.1.7}$$

We need f defined only on $\mathcal{F} \times \emptyset$ to define $f^{\wedge+}(z)$, and only on $\mathcal{F} \times (\mathcal{F}_d \setminus \{\emptyset\})$ to define $f^{\wedge-}(z,\zeta)$.

THEOREM 4.1.1. *Let $\Sigma(U)$ be a system with time axis $\mathcal{F}_d \times \mathcal{F}_d$ as in* (4.1.2)-(4.1.3). *Then the map $\mathcal{I}'$ defined by* (4.1.5) *is a bijection between $\mathcal{T}$ and* $\ell(\mathcal{F}_d \times$

$\{\emptyset\}, \mathcal{E}) \times \mathcal{H} \times \ell(\mathcal{F}_d \times (\mathcal{F}_d \setminus \{\emptyset\}), \mathcal{E}_*)$. *Explicitly, given* $(u_+, h, y_-) \in \ell(\mathcal{F}_d \times \{\emptyset\}) \times \mathcal{H} \times \ell(\mathcal{F}_d \times (\mathcal{F}_d \setminus \{\emptyset\}), \mathcal{E}_*)$, $(u, x, y) = \mathcal{I}'^{-1}(u_+, h, y_-)$ *is determined by*

$$\begin{aligned}\widehat{u}(z,\zeta) &= (I + B^* Z_r(\zeta)^*(I - A^* Z_r(\zeta)^*)^{-1}(I - Z_r(z)A)^{-1} Z_r(z)B) u_+^{\wedge+}(z)\\ &\quad + B^* Z_r(\zeta)^*(I - A^* Z_r(\zeta)^*)^{-1}(I - Z_r(z)A)^{-1} h\\ &\quad + T_{\Sigma(U)}(\zeta)^* y_-^{\wedge-}(z,\zeta) \end{aligned} \tag{4.1.8}$$

$$\begin{aligned}\widehat{x}(z,\zeta) &= (I - A^* Z_r(\zeta)^*)^{-1}(I - Z_r(z)A)^{-1}\left[h + Z_r(z)Bu_+^{\wedge+}(z)\right]\\ &\quad + (I - A^* Z_r(\zeta)^*)^{-1} C^* y_-^{\wedge-}(z,\zeta)\end{aligned} \tag{4.1.9}$$

$$\widehat{y}(z,\zeta) = y_-^{\wedge-}(z,\zeta) + C(I - Z_r(z)A)^{-1} h + T_{\Sigma(U)}(z) u_+^{\wedge+}(z), \tag{4.1.10}$$

where

$$\begin{aligned} T_{\Sigma(U)}(z) &= D + C(I - Z_r(z)A)^{-1} Z_r(z) B\\ &= D + \sum_{k=1}^{d} \sum_{v \in \mathcal{F}_d} C A^v B_k z^v z_k \end{aligned} \tag{4.1.11}$$

is the transfer function *of the system* $\Sigma(U)$, *also called the* characteristic function *of the colligation* U, *where*

$$T_{\Sigma(U)}(\zeta)^* = D^* + B^* Z_r(\zeta)^*(I - A^* Z_r(\zeta)^*)^{-1} C^* \tag{4.1.12}$$

$$= D^* + \sum_{k=1}^{d} \sum_{w \in \mathcal{F}_d} B_k^* A_k^{*w} C^* \zeta_k \zeta^w \tag{4.1.13}$$

is the adjoint transfer function *for* $\Sigma(U)$, *and where we have set*

$$Z_r(z) = \begin{bmatrix} z_1 I_{\mathcal{H}} & \cdots & z_d I_{\mathcal{H}} \end{bmatrix}, \qquad Z_r(\zeta)^* = \begin{bmatrix} \zeta_1 I_{\mathcal{H}} \\ \vdots \\ \zeta_d I_{\mathcal{H}} \end{bmatrix}.$$

PROOF. Assume first that (u, x, y) is a system trajectory. Multiplication of the first equation in (4.1.2) by $z_j z^v$ gives

$$x(g_j v, \emptyset) z_j z^v = (z_j A_j) x(v, \emptyset) z^v + (z_j B_j) u(v, \emptyset) z^v \text{ for } j = 1, \ldots, d.$$

Summing these over $j = 1, \ldots, d$ and all $v \in \mathcal{F}_d$ gives

$$x^{\wedge+}(z) - x(\emptyset, \emptyset) z^{\emptyset} = Z_r(z) A x^{\wedge+}(z) + Z_r(z) B u^{\wedge+}(z).$$

Solving for $x^{\wedge+}(z)$ then gives

$$x^{\wedge+}(z) = (I - Z_r(z)A)^{-1} x(\emptyset, \emptyset) + (I - Z_r(z)A)^{-1} Z_r(z) B u^{\wedge+}(z). \tag{4.1.14}$$

Applying the forward Z-transform to the second equation in (4.1.2) then gives

$$\begin{aligned} y^{\wedge+}(z) &= C x^{\wedge+}(z) + D u^{\wedge+}(z)\\ &= C(I - Z_r(z)A)^{-1} x(\emptyset, \emptyset) + C(I - Z_r(z)A)^{-1} Z_r(z) B u^{\wedge+}(z) + D u^{\wedge+}(z) \end{aligned} \tag{4.1.15}$$

and hence

$$\begin{aligned} \widehat{y}(z) &= y^{\wedge-}(z,\zeta) + y^{\wedge+}(z)\\ &= y^{\wedge-}(z,\zeta) + C(I - Z_r(z)A)^{-1} x(\emptyset, \emptyset) + T_{\Sigma(U)}(z) u^{\wedge+}(z) \end{aligned}$$

which is exactly (4.1.10) if $(u_+, h, y_-) = \mathcal{I}'(u, x, y)$. Multiplying (4.1.3) by $z^v \zeta_j \zeta^w$ gives

$$x(v, g_j w) z^v \zeta_j \zeta^w = (A_j^* \zeta_j) x(v, w) z^v \zeta^w + C^* y(v, g_j w) z^v \zeta_j \zeta^w$$
$$u(v, g_j w) z^v \zeta_j \zeta^w = (B_j^* \zeta_j) x(v, w) z^v \zeta^w + D^* y(v, g_j w) z^v \zeta_j \zeta^w.$$

Summing over all $v, w \in \mathcal{F}_d$ and all $j = 1, \ldots, d$ gives

$$\widehat{x}(z, \zeta) - x^{\wedge +}(z) = A^* Z_r(\zeta)^* \widehat{x}(z, \zeta) + C^* y^{\wedge -}(z, \zeta) \tag{4.1.16}$$

$$u^{\wedge -}(z, \zeta) = B^* Z_r(\zeta)^* \widehat{x}(z, \zeta) + D^* y^{\wedge -}(z, \zeta). \tag{4.1.17}$$

Solving (4.1.16) for $\widehat{x}(z, \zeta)$ gives

$$\widehat{x}(z, \zeta) = (I - A^* Z_r(\zeta)^*)^{-1} x^{\wedge +}(z) + (I - A^* Z_r(\zeta)^*)^{-1} C^* y^{\wedge -}(z, \zeta). \tag{4.1.18}$$

Substituting (4.1.14) for $x^{\wedge +}(z)$ in (4.1.18) gives

$$\begin{aligned} \widehat{x}(z, \zeta) = {} & (I - A^* Z_r(\zeta)^*)^{-1} (I - Z_r(z) A)^{-1} [x(\emptyset, \emptyset) + Z_r(z) B u^{\wedge +}(z)] \\ & + (I - A^* Z_r(\zeta)^*)^{-1} C^* y^{\wedge -}(z, \zeta) \end{aligned} \tag{4.1.19}$$

which is exactly (4.1.9) if $(u_+, h, y_-) = \mathcal{I}'(u, x, y)$. Finally, substituting (4.1.19) for $\widehat{x}(z, \zeta)$ in (4.1.17) gives

$$\begin{aligned} u^{\wedge -}(z, \zeta) = {} & B^* Z_r(\zeta)^* (I - A^* Z_r(\zeta)^*)^{-1} (I - Z_r(z) A)^{-1} h \\ & + B^* Z_r(\zeta)^* (I - A^* Z_r(\zeta)^*)^{-1} (I - Z_r(z) A)^{-1} Z_r(z) B u^{\wedge +}(z) \\ & + T_{\Sigma(U)}(\zeta)^* y^{\wedge -}(z, \zeta). \end{aligned}$$

Thus

$$\begin{aligned} \widehat{u}(z, \zeta) & = u^{\wedge +}(z) + u^{\wedge -}(z, \zeta) \\ & = (I + B^* Z_r(\zeta)^* (I - A^* Z_r(\zeta)^*)^{-1} (I - Z_r(z) A)^{-1} Z_r(z) B) u^{\wedge +}(z) \\ & \quad + B^* Z_r(\zeta)^* (I - A^* Z_r(\zeta)^*)^{-1} (I - Z_r(z) A)^{-1} h \\ & \quad + T_{\Sigma(U)}(\zeta)^* y^{\wedge -}(z, \zeta) \end{aligned} \tag{4.1.20}$$

which is exactly (4.1.8) if $(u_+, h, y_-) = \mathcal{I}'(u, x, y)$. Thus (4.1.8)–(4.1.10) holds whenever $(u_+, h, y_-) = \mathcal{I}'(u, x, y)$.

Conversely, given $(u_+, h, y_-) \in \ell(\mathcal{F}_d, \mathcal{E}) \times \mathcal{H} \times \ell(\mathcal{F}_d \times (\mathcal{F}_d \setminus \{\emptyset\}), \mathcal{E}_*)$, define (u, x, y) by (4.1.8)–(4.1.10). Then clearly $\mathcal{I}'(u, x, y) = (u_+, h, y_-)$ and it is a direct verification that (u, x, y) satisfies the frequency-domain version of the system equations (4.1.8)–(4.1.10), namely:

$$\begin{aligned} z_j^{-1} x^{\wedge +}(z) & = A_j x^{\wedge +}(z) + B_j u^{\wedge +}(z) \\ y^{\wedge +}(z) & = C x^{\wedge +}(z) + D u^{\wedge +}(z) \\ \zeta_j^{-1} \widehat{x}(z, \zeta) & = A_j^* \widehat{x}(z, \zeta) + C^* \zeta_j^{-1} \widehat{y}(z, \zeta) \\ \zeta_j^{-1} \widehat{u}(z, \zeta) & = B_j^* \widehat{x}(z, \zeta) + D^* \zeta_j^{-1} \widehat{y}(z, \zeta). \end{aligned} \tag{4.1.21}$$

This completes the proof of Theorem 4.1.1. □

Now let us suppose that U is contractive. Then we have the dissipation inequality

$$\sum_{k=1}^{d} \|x(g_k v, \emptyset)\|^2 - \|x(v, \emptyset)\|^2 \leq \|u(v, \emptyset)\|^2 - \|y(v, \emptyset)\|^2 \tag{4.1.22}$$

over all trajectories of the system $\Sigma(U)$. If we take the sum of both sides of (4.1.22) over all words v of length at most N, we get

$$\sum_{v\colon |v|=N+1} \|x(v,\emptyset)\|^2 - \|x(\emptyset,\emptyset)\|^2 \leq \sum_{v\colon |v|\leq N} \left[\|u(v,\emptyset)\|^2 - \|y(v,\emptyset)\|^2\right]. \tag{4.1.23}$$

If we assume that $u|_{\mathcal{F}_d\times\{\emptyset\}} \in \ell^2(\mathcal{F}_d\times\{\emptyset\},\mathcal{E})$, we see that

$$\begin{aligned} \|y|_{\mathcal{F}_d\times\{\emptyset\}}\|^2_{\ell^2(\mathcal{F}_d,\mathcal{E}_*)} &\leq \|u|_{\mathcal{F}_d\times\{\emptyset\}}\|^2_{\ell^2(\mathcal{F}_d,\mathcal{E})} + \|x(\emptyset,\emptyset)\|^2 \\ &\qquad - \lim_{N\to\infty}\textstyle\sum_{v\colon |v|=N+1}\|x(v,\emptyset)\|^2 \\ &\leq \|u|_{\mathcal{F}_d\times\{\emptyset\}}\|^2_{\ell^2(\mathcal{F}_d,\mathcal{E})} + \|x(\emptyset,\emptyset)\|^2. \end{aligned} \tag{4.1.24}$$

If U is isometric and if the stability condition

$$\lim_{N\to\infty} \sum_{v\colon |v|=N+1} \|x(v,\emptyset)\|^2 = 0 \tag{4.1.25}$$

holds, then in fact we have equality throughout in (4.1.23) and (4.1.24). The content of the next lemma is that (4.1.25) holding for all trajectories of the system with norm-square summable input signal u is equivalent to the d-tuple $A=(A_1,\dots,A_d)$ being *d-stable* in the sense that

$$\lim_{N\to\infty}\left\{\sum_{v\colon |v|=N}\|A^v h\|^2\right\} = 0 \text{ for each } h\in\mathcal{H}. \tag{4.1.26}$$

In the statement and proof of this lemma and the proposition immediately following, we are considering only forward trajectories, so we identify $\mathcal{F}_d\times\{\emptyset\}$ with $\mathcal{F}_d$ and suppress the second argument $\emptyset$ in the trajectory functions u,x,y.

LEMMA 4.1.2. *Suppose that U as in (4.1.1) is a d-variable, contractive colligation with associated forward system equations*

$$\Sigma_f\colon \left\{\begin{array}{rcl} x(g_jv) &=& A_jx(v)+B_ju(v) \text{ for } j=1,\dots,d \\ y(v) &=& Cx(v)+Du(v). \end{array}\right. \tag{4.1.27}$$

and assume that A is d-stable as defined in (4.1.26). Let (u,x,y) be any solution of (4.1.27) with $u\in\ell^2(\mathcal{F}_d,\mathcal{E})$. Then x satisfies the stability condition (4.1.25).

PROOF. Let (u,x,y) be as in the statement of the lemma and let $\epsilon>0$. We must find $M<\infty$ so that

$$n>M \Longrightarrow \sum_{v\in\mathcal{F}_d\colon |v|=n}\|x(n)\|^2<\epsilon. \tag{4.1.28}$$

Since

$$\|u\|^2=\sum_{n=0}^{\infty}\left[\sum_{v\in\mathcal{F}_d\colon |v|=n}\|u(v)\|^2\right]<\infty,$$

as a first step we choose $K<\infty$ so that

$$\sum_{n=K}^{\infty}\left[\sum_{v\in\mathcal{F}_d\colon |v|=n}\|u(v)\|^2\right]<\epsilon/3. \tag{4.1.29}$$

We next decompose $u\in\ell^2(\mathcal{F}_d,\mathcal{E})$ as $u=u_1+u_2$ where

$$u_1(v):=\begin{cases} u(v) & \text{if } |v|<K, \\ 0 & \text{if } |v|\geq K \end{cases}, \qquad u_2(v):=\begin{cases} 0 & \text{if } |v|<K, \\ u(v) & \text{if } |v|\geq K \end{cases}.$$

Then by linearity we may write $x(v) = x_0(v) + x_1(v) + x_2(v)$ where x_0 is the state trajectory for (4.1.27) associated with initial condition $x_0(\emptyset) = x(\emptyset)$ and with zero input signal, and where, for $j = 1, 2$, x_j is the state trajectory for (4.1.27) with initial condition $x_j(\emptyset) = 0$ and with input signal equal to u_j. To show that (4.1.28) holds, it suffices to show that, for each $j = 0, 1, 2$, there is a $M_j < \infty$ so that

$$n > M_j \Longrightarrow \sum_{v \in \mathcal{F}_d \colon |v| = n} \|x_j(v)\|^2 < \epsilon/3. \tag{4.1.30}$$

Indeed, once we have obtained (4.1.30), we simply take $M = \max\{M_0, M_1, M_3\}$ to verify (4.1.28).

Consider first the case $j = 0$. From the formula (4.1.9) we read off that $x_0(v) = A^v x(\emptyset)$ and hence $\sum_{v\colon |v|=n} \|x_0(v)\|^2 = \sum_{v\colon |v|=n} \|A^v x(\emptyset)\|^2$. Hence the hypothesis (4.1.26) implies the existence of an $M_0 < \infty$ for which (4.1.30) holds for $j = 0$.

Consider next the case $j = 1$. For $v \in \mathcal{F}_d$ with $|v| > K$, write $v = v''v'$ where $|v'| = K$. Then we are in the situation of the $j = 0$ case if we consider the system initialized at v' rather than at $\emptyset$; we get $x_1(v''v') = A^{v''} x_1(v')$. The hypothesis (4.1.26) implies the existence of an $M_{1,v'} < \infty$ so that

$$n > M_{1,v'} \Longrightarrow \sum_{v''\colon |v''|=n} \|x_1(v''v')\|^2 < \epsilon/3d^K.$$

If we set $M_1 = \sup\{K + M_{1,v'} \colon v' \in \mathcal{F}_d \text{ with } |v'| = K\}$, then M_1 satisfies (4.1.30) for $j = 1$.

Finally consider the case $j = 2$. From (4.1.23) we see that

$$\sum_{v\colon |v|=n} \|x_2(v)\|^2 \leq \|u_2\|^2_{L^2} < \epsilon/3$$

for all $n = 1, 2, 3, \ldots$ by the way u_2 was constructed. Hence (4.1.30) is satisfied for $j = 2$ with $M_2 = 1$. This completes the proof of the lemma. $\square$

The discussion preceding Lemma 4.1.2 combined with the result of the lemma leads to the following result.

PROPOSITION 4.1.3. *Let U be a d-variable, contractive colligation with associated contractive linear system Σ. Then:*

(1) *The observability operator*

$$\mathcal{O}\colon x \mapsto C(I - Z_r(z)A)^{-1}x \text{ from } \mathcal{H} \text{ into } L^2(\mathcal{F}_d, \mathcal{E}_*)$$

is contractive. If U is a d-variable isometric colligation and the operator-tuple $A = (A_1, \ldots, A_d)$ is d-stable, then $\mathcal{O}\colon \mathcal{H} \to L^2(\mathcal{F}_d, \mathcal{E}_*)$ is isometric.

(2) *The input-output operator*

$$\widehat{u}(z) \mapsto T_{\Sigma(U)}(z) \cdot \widehat{u}(z) \text{ from } L^2(\mathcal{F}_d, \mathcal{E}) \text{ to } L^2(\mathcal{F}_d, \mathcal{E}_*)$$

is contractive, i.e., $T_{\Sigma(U)}(z) \in \mathcal{S}_{nc,d}(\mathcal{E}, \mathcal{E}_*)$. If U is isometric and $A = (A_1, \ldots, A_d)$ is d-stable, then

$$T_{\Sigma(U)}\colon L^2(\mathcal{F}_d, \mathcal{E}) \to L^2(\mathcal{F}_d, \mathcal{E}_*)$$

(as a formal left multiplication operator) is isometric.

PROOF. If U is only assumed to be contractive, then the first statement of the Proposition 4.1.3 follows by applying the inequality (4.1.24) for the special case $u = 0$ while the second statement follows from the special case $x(\emptyset) = 0$.

In case U is isometric and A is d-stable, by the discussion preceding the lemma we see that (4.1.23) holds with equality for all trajectories (u, x, y) of the system:

$$\sum_{v\colon |v|=N+1} \|x(v)\|^2 + \sum_{v\colon |v|\le N} \|y(v)\|^2 = \|x(\emptyset)\|^2 + \sum_{v\colon |v|\le N} \|u(v)\|^2. \tag{4.1.31}$$

By the result of Lemma (4.1.2) we know that the first term on the left goes to zero as $n \to \infty$. Hence taking the limit as $N \to \infty$ in (4.1.31) leaves us with

$$\|y\|^2_{\ell^2(\mathcal{F}_d, \mathcal{E}_*)} = \|x(\emptyset)\|^2 + \|u\|^2_{\ell^2(\mathcal{F}_d, \mathcal{E})}. \tag{4.1.32}$$

Application of (4.1.32) for the special case $u = 0$ implies that $\mathcal{O}$ is isometric, and the special case $x(\emptyset) = 0$ gives that multiplication by $T_{\Sigma(U)}$ is isometric from $L^2(\mathcal{F}_d, \mathcal{E})$ into $L^2(\mathcal{F}_d, \mathcal{E}_*)$. This concludes the proof. □

4.2. The Cuntz scattering system associated with a unitary colligation U and Cuntz weight W_*

Let U be a d-variable unitary colligation as in (4.1.1). Suppose in addition that we are given a fixed Cuntz-weight extension W_* of the identity with matrix entries equal to operators on the output space $\mathcal{E}_*$, so

$$W_{*v,\emptyset;\alpha,\emptyset} = \delta_{v,\alpha} I_{\mathcal{E}_*}.$$

Define the space $\mathcal{T}_{U,W_*}$ of *admissible trajectories* of $\Sigma(U)$ to consist of all solutions (u, x, y) of the system equations (4.1.2) and (4.1.3) for which

$$u|_{\mathcal{F}_d \times \{\emptyset\}} \in \ell^2(\mathcal{F}_d, \mathcal{E}), \qquad y \in \ell_{W_*}. \tag{4.2.1}$$

Note that if U is the unitary colligation $U(\mathfrak{S})$ coming from a Cuntz scattering system as in Theorem 3.3.1, and if W_* is the Cuntz weight associated with the incoming wandering subspace $\mathcal{E}_*$, then $\mathcal{T}_{U,W_*} = \operatorname{im}\Omega$ where Ω is defined as in (3.3.18), by Theorem 3.3.7. By Theorem 3.3.8 the norm of an element $k \in \mathcal{K}$ can be expressed in terms of its image $(u, x, y) = \Omega k$ as

$$\|(u, x, y)\|^2_{\mathcal{T}_{U,W_*}} = \|P_{h^\perp_{W_*}} y\|^2_{h^\perp_{W_*}} + \|x(\emptyset, \emptyset)\|^2_{\mathcal{H}} + \|u|_{\mathcal{F}_d \times \{\emptyset\}}\|^2_{\ell^2(\mathcal{F}_d, \mathcal{E})}. \tag{4.2.2}$$

Moreover, from (3.3.23) we see that the action of $\mathcal{U}_j$ on $\mathcal{K}$ corresponds to the bilateral shift $(u, x, y) \mapsto (\widetilde{U}^R_j u, \widetilde{U}^R_j x, \widetilde{U}^R_j y)$ on $\operatorname{im}\Omega = \mathcal{T}_{U,W}$. Our goal here is to get back to a Cuntz scattering system $\mathfrak{S} = \mathfrak{S}(\Sigma(U), W_*)$ starting from a d-variable unitary system $\Sigma(U)$ and a choice of shift Cuntz weight W_* on the output space $\mathcal{E}_*$ for the colligation U.

Let us therefore begin with a Cuntz unitary colligation U and (4.1.1) together with shift Cuntz weight W_* with matrix entries acting on $\mathcal{E}_*$. It is natural to define a map $\mathcal{I}_{U,W_*} \colon \mathcal{T}_{U,W_*} \to \mathcal{H}^\perp_{W_*} \oplus \mathcal{H} \oplus L^2(\mathcal{F}_d, \mathcal{E})$ by

$$\mathcal{I}_{U,W_*} \colon (u, x, y) \mapsto \begin{bmatrix} (P_{\mathcal{H}^\perp_{W_*}} \widehat{y})(z, \zeta) \\ x(\emptyset, \emptyset) \\ u^{\wedge +}(z) \end{bmatrix} \tag{4.2.3}$$

where, as usual, we set $\widehat{y}(z, \zeta) = \sum_{v,w \in \mathcal{F}_d} y(v, w) z^v \zeta^w$ equal to the noncommutative Z-transform of y. The next result gives the basic properties of the map $\mathcal{I}_{U,W_*}$.

LEMMA 4.2.1. *The map $\mathcal{I}_{U,W_*} : \mathcal{T}_{U,W_*} \to \mathcal{H}_{W_*}^{\perp} \oplus \mathcal{H} \oplus L^2(\mathcal{F}_d, \mathcal{E})$ defined by* (4.2.3) *is bijective with inverse given by*

$$(\mathcal{I}_{U,W_*})^{-1} \colon \begin{bmatrix} y_\perp \\ h \\ u_+ \end{bmatrix} \mapsto (u, x, y)$$

with (u, x, y) given by

$$\begin{aligned} \widehat{u}(z,\zeta) &= \widehat{W}(z,\zeta)u_+^{\wedge +}(z) + B^* Z_r(\zeta)^*(I - A^* Z_r(\zeta)^*)^{-1}(I - Z_r(z)A)^{-1}h \\ &\quad + T(\zeta)^*(\widehat{W}_*(z,\zeta) - I)C(I - Z_r(z)A)^{-1}h + T(\zeta)^* y_\perp^{\wedge -}(z,\zeta), \\ \widehat{x}(z,\zeta) &= (I - A^* Z_r(\zeta)^*)^{-1}(I - Z_r(z)A)^{-1}h \\ &\quad + (I - A^* Z_r(\zeta)^*)^{-1} C^*(\widehat{W}_*(z,\zeta) - I)C(I - Z_r(z)A)^{-1}h \\ &\quad + (I - A^* Z_r(\zeta)^*)^{-1}(I - Z_r(z)A)^{-1} Z_r(z) B u_+^{\wedge +}(z) \\ &\quad + (I - A^* Z_r(\zeta)^*)^{-1} C^*(W_*(z,\zeta) - I)T(z)u_+^{\wedge +}(z) \\ &\quad + (I - A^* Z_r(\zeta)^*)^{-1} C^* y_\perp^{\wedge -}(z,\zeta), \end{aligned}$$

$$\widehat{y}(z,\zeta) = y_\perp^{\wedge -}(z,\zeta) + \widehat{W}_*(z,\zeta)\left[C(I - Z_r(z)A)^{-1}h + T(z)u_+^{\wedge +}(z)\right]. \tag{4.2.4}$$

where $\widehat{W}$ is given by

$$\begin{aligned} \widehat{W}(z,\zeta) =& I + B^* Z_r(\zeta)^*(I - A^* Z_r(\zeta)^*)^{-1}(I - Z_r(z)A)^{-1} Z_r(z) B \\ & + T(\zeta)^*(\widehat{W}_*(z,\zeta) - I)T(z) \end{aligned} \tag{4.2.5}$$

and where we have set $T(z) = T_{\Sigma(U)}(z)$ equal to the characteristic function for the colligation U (see (4.1.11) *and* (4.1.13)*). In particular, $\mathcal{T}_{U,W_*}$ is a Hilbert space in the norm* (4.2.2).

PROOF. To show that $\mathcal{I}_{U,W_*}$ is onto, given $y_\perp \oplus h \oplus u_+ \in h_{W_*}^{\perp} \oplus \mathcal{H} \oplus \ell^2(\mathcal{F}_d, \mathcal{E})$, we must produce a $(u, x, y) \in \mathcal{T}_{U,W_*}$ so that $\mathcal{I}_{W_*}(u, x, y) = y_\perp \oplus h \oplus u_+$.

From the formulas (4.1.8), (4.1.9) and (4.1.10) we see that the forward parts of (u, x, y) are forced to be

$$\begin{aligned} u^{\wedge +}(z) = u_+^{\wedge +}(z),\ x^{\wedge +}(z) &= (I - Z_r(z)A)^{-1}h + (I - Z_r(z)A)^{-1} Z_r(z) B u_+^{\wedge +}(z), \\ y^{\wedge +}(z) &= C(I - Z_r(z)A)^{-1}h + T(z)u_+^{\wedge +}(z). \end{aligned}$$

and we also require that

$$(P_{\mathcal{H}_{W_*}^{\perp}} \widehat{y})(z,\zeta) = \widehat{y_\perp}(z,\zeta).$$

Since W_* is a shift Cuntz weight, we have that the coefficients of $P_{\mathcal{H}_{W_*}^{\perp}} \widehat{y}$ are supported on $\mathcal{F}_d \times (\mathcal{F}_d \setminus \{\emptyset\})$ and hence $\widehat{y_\perp}(z,\zeta) = y_\perp^{\wedge -}(z,\zeta)$. By Proposition 2.3.3 we know that $P_{\mathcal{H}_{W_*}} \widehat{y}$ is uniquely determined by $y^{\wedge +}$ according to the formula

$$(P_{\mathcal{H}_{W_*}} \widehat{y})(z,\zeta) = \widehat{W}_*(z,\zeta) y^{\wedge +}(z).$$

We conclude that $\widehat{y}(z,\zeta)$ must be given by

$$\begin{aligned} \widehat{y}(z,\zeta) &= (P_{\mathcal{H}_{W_*}^{\perp}} \widehat{y})(z,\zeta) + (P_{\mathcal{H}_{W_*}} \widehat{y})(z,\zeta) \\ &= y_\perp^{\wedge -}(z,\zeta) + \widehat{W}_*(z,\zeta)\left[C(I - Z_r(z)A)^{-1}h + T(z)u_+^{\wedge +}(z)\right]. \end{aligned}$$

Therefore, since W_* by assumption is a shift Cuntz weight, the part of y supported in the past is given by

$$(4.2.6)\ y^{\wedge -}(z,\zeta) = y_{\perp}^{\wedge -}(z,\zeta) + (\widehat{W}_*(z,\zeta) - I)\left[C(I - Z_r(z)A)^{-1}h + T(z)u_+^{\wedge +}(z)\right].$$

We now have identified the free parameters $u|_{\mathcal{F}_d \times \{\emptyset\}}$, $x(\emptyset,\emptyset)$ and $y|_{\mathcal{F}_d \times (\mathcal{F}_d \setminus \{\emptyset\})}$ needed to determine the complete trajectory (u, x, y) according to the prescription (4.1.8), (4.1.9), (4.1.10) in Theorem 4.1.1, namely:

$$u^{\wedge +}(z) = u_+^{\wedge +}(z), \qquad x(\emptyset,\emptyset) = h, \quad y^{\wedge -}(z,\zeta) \text{ as in (4.2.6).}$$

Applying the formulas (4.1.8)–(4.1.10) then gives us the formula (4.2.4) for the operator $(\mathcal{I}_{U,W_*})^{-1}$. Injectivity follows from the uniqueness in Theorem 4.1.1. As $\|(u,x,y)\|^2_{\mathcal{T}_{U,W_*}} = \|\mathcal{I}_{U,W_*}(u,x,y)\|^2_{\mathcal{H}^{\perp}_{W_*} \oplus \mathcal{H} \oplus \ell^2(\mathcal{F}_d,\mathcal{E})}$, it follows that $\mathcal{T}_{U,W_*}$ is a Hilbert space in the $\mathcal{T}_{U,W_*}$-norm given by (4.2.2). □

The next result is another step toward understanding the equivalence of a pair of the form (U, W_*) versus a Cuntz scattering system $\mathfrak{S}$.

PROPOSITION 4.2.2. *Let U be a Cuntz unitary colligation* (4.1.1) *and W_* a Cuntz shift weight with matrix entries equal to operators acting on the output space $\mathcal{E}_*$ of U. Let $T(z) = T_{\Sigma(U)}(z)$ denote the transfer function* (4.1.11) *for the system* (4.1.2)–(4.1.3) *(also referred to as the characteristic function of the colligation U), and let W be the* [∗]*-Haplitz operator with symbol $\widehat{W}$ given by* (4.2.5)*. Then:*

1. *W is a shift Cuntz weight and (T, W, W_*) is admissible in the sense of Theorem 2.4.5, i.e., the formula*
$$S\colon W[p] \mapsto W_* L_T[p] \ \textit{for}\ p \in L^2(\mathcal{F}_d \times \mathcal{F}_d, \mathcal{E})$$
extends by continuity to define a contraction operator $S : L_T^{W,W_*} : \mathcal{L}_W \to \mathcal{L}_{W_*}$.
2. *In case $U = U(\mathfrak{S})$ is the Cuntz unitary colligation arising form the scattering system $\mathfrak{S}$ as in* (3.3.2) *with outgoing Cuntz weight W_*, incoming Cuntz weight W and scattering function $S\colon \mathcal{L}_W \to \mathcal{L}_{W_*}$, then the symbol $\widehat{W}$ for W is given by* (4.2.5) *and S is given by $S = L_T^{W,W_*}$. In particular, W and S are completely determined by the pair (U, W_*).*

PROOF. From the formula (4.2.5) for $\widehat{W}(z,\zeta)$, we see that W is a Haplitz extension of the identity whenever W_* is. Thus, to verify part (1) of the Theorem, it remains only to verify that (T, W, W_*) is admissible in the sense of Theorem 2.4.5. By part (2) of that theorem, this assertion follows once we check that

$$(4.2.7)\qquad X(z,\zeta) := \widehat{W}(z,\zeta) - I - T(\zeta)^*[\widehat{W}_*(z,\zeta) - I]T(z)$$

is a positive symbol for which $X'(z,\zeta) = 0$, where

$$(4.2.8)\qquad X'(z,\zeta) := X(z,\zeta) - \sum_{k=1}^{d} z_k^{-1} X(z,\zeta)\zeta_k^{-1} + I - T(\zeta)^*T(z).$$

Substituting the formula (4.2.5) for $\widehat{W}(z,\zeta)$ in the formula (4.2.7) gives

$$(4.2.9)\qquad X(z,\zeta) = B^* Z_r(\zeta)^*(I - A^* Z_r(\zeta)^*)^{-1}(I - Z_r(z)A)^{-1}Z_r(z)B.$$

From this factored form for $X(z,\zeta)$ it is clear that $X(z,\zeta)$ is symbol-positive. Recalling the formula $T(z) = D + C(I - Z_r(z)A)^{-1}Z_r(z)B$ for $T(z)$ and using the identities

$$I - D^*D = B^*B, \qquad -D^*C = B^*A, \qquad -C^*D = A^*B, \qquad C^*C = -A^*A + I$$

arising out of the fact that U is unitary, we compute

$$\begin{aligned} I - T(\zeta)^*T(z) &= I - D^*D - D^*C(I - Z_r(z)A)^{-1}Z_r(z)B \\ &\quad - B^*Z_r(\zeta)^*(I - A^*Z(\zeta)^*)^{-1}C^*D \\ &\quad - B^*Z_r(\zeta)^*(I - A^*Z_r(\zeta)^*)^{-1}C^*C(I - Z_r(z)A)^{-1}Z_r(z)B \\ &= B^*B + B^*A(I - Z_r(z)A)^{-1}Z_r(z)B + B^*Z_r(\zeta)^*(I - A^*Z_r(\zeta)^*)^{-1}A^*B \\ &\quad + B^*Z_r(\zeta)^*(I - A^*Z_r(\zeta)^*)^{-1}A^*A(I - Z_r(z)A)^{-1}Z_r(z)B - X(z,\zeta). \end{aligned}$$

From the expression (4.2.8) for $X'(z,\zeta)$ we see that the demand that $X'(z,\zeta) = 0$ comes down to

$$\begin{aligned} &\sum_{k=1}^{d} z_k^{-1}\left(B^*Z_r(\zeta)^*(I - A^*Z_r(\zeta)^*)^{-1}(I - Z_r(z)A)^{-1}Z_r(z)B\right)\zeta_k^{-1} \\ &\qquad = B^*B + B^*A(I - Z_r(z)A)^{-1}Z_r(z)B + B^*Z_r(\zeta)^*(I - A^*Z_r(\zeta)^*)^{-1}A^*B \\ &\qquad\quad + B^*Z_r(\zeta)^*(I - A^*Z_r(\zeta)^*)^{-1}A^*A(I - Z_r(z)A)^{-1}Z_r(z)B, \end{aligned} \tag{4.2.10}$$

or, directly in terms of the coefficients of formal power series,

$$\begin{aligned} &\sum_{k=1}^{d} z_k^{-1}\left[\left(\sum_{j=1}^{d}\zeta_j B_j^*\right)\left(\sum_{w\in\mathcal{F}_d} A^{*w}\zeta^{w}\right)\cdot\left(\sum_{v\in\mathcal{F}_d} A^{v}z^{v}\right)\cdot\left(\sum_{\ell=1}^{d} B_\ell z_\ell\right)\right]\zeta_k^{-1} \\ &\quad = \left[\sum_{j=1}^{d} B_j^*B_j\right] + \left[\left(\sum_{j=1}^{d} B_j^*A_j\right)\cdot\left(\sum_{v\in\mathcal{F}_d} A^{v}z^{v}\right)\cdot\left(\sum_{\ell=1}^{d} B_\ell z_\ell\right)\right] \\ &\qquad + \left[\left(\sum_{j=1}^{d} B_j^*\zeta_j\right)\cdot\left(\sum_{w\in\mathcal{F}_d} A^{*w}\zeta^{w}\right)\cdot\left(\sum_{\ell=1}^{d} A_\ell^*B_\ell\right)\right] \\ &\qquad + \left[\left(\sum_{j=1}^{d} B_j^*\zeta_j\right)\cdot\left(\sum_{w\in\mathcal{F}_d} A^{*w}\zeta^{w}\right)\cdot A^*A\cdot\left(\sum_{v\in\mathcal{F}_d} A^{v}z^{v}\right)\cdot\left(\sum_{\ell=1}^{d} B_\ell z_\ell\right)\right]. \end{aligned}$$

Upon summation over $k = 1,\dots,d$, this in turn follows from verifying that

$$\begin{aligned} &z_k^{-1}\cdot\left[\left(\sum_{j=1}^{d} B_j^*\zeta_j\right)\cdot\left(\sum_{w\in\mathcal{F}_d} A^{*w}\zeta^{w}\right)\cdot\left(\sum_{v\in\mathcal{F}_d} A^{v}z^{v}\right)\cdot\left(\sum_{\ell=1}^{d} B_\ell z_\ell\right)\right]\cdot\zeta_k^{-1} \\ &\quad = B_k^*B_k + \left[\left(\sum_{j=1}^{d} B_j^*\zeta_j\right)\cdot\left(\sum_{w\in\mathcal{F}_d} A^{*w}\zeta^{w}\right)\cdot A_k^*B_k\right] \\ &\qquad + \left[B_k^*A_k\cdot\left(\sum_{v\in\mathcal{F}_d} A^{v}z^{v}\right)\cdot\left(\sum_{\ell=1}^{d} B_\ell z_\ell\right)\right] \end{aligned}$$

$$+\left[\left(\sum_{j=1}^{d} B_j^* \zeta_j\right) \cdot \left(\sum_{w \in \mathcal{F}_d} A^{*w} \zeta^w\right) \cdot A_k^* A_k \cdot \left(\sum_{v \in \mathcal{F}_d} A^v z^v\right) \left(\sum_{\ell=1}^{d} B_\ell z_\ell\right)\right]$$

for each $k = 1, \dots, d$. Verification of this last equality is now an elementary exercise in the algebra of noncommutative formal power series which we leave to the reader. With this verification of (4.2.10), the proof of part (1) of Proposition 4.2.2 is complete.

Suppose now that $U = U(\mathfrak{S})$ (as in (3.3.2)) for a scattering system with incoming Cuntz weight W_*, outgoing Cuntz weight W and scattering operator S. Temporarily denote the Haplitz operator with symbol given by the right-hand side of (4.2.5) by W' rather than by W. By Theorem 3.3.7 we know that $\operatorname{im}\Omega$ is characterized as the set of those trajectories (u, x, y) for $\Sigma(U)$ (see (4.1.2)–(4.1.3)) for which $u \in \ell_W$ and $y \in \ell_{W_*}$. As $u^{\wedge+} \in L^2(\mathcal{F}_d, \mathcal{E})$ whenever $u \in \ell_W$ since W is a Cuntz weight, we immediately see the containment

$$\operatorname{im}\Omega \subset \mathcal{T}_{U,W_*}. \tag{4.2.11}$$

Moreover, from the definitions we see that

$$\mathcal{I}_{U,W_*}\Omega\mathcal{G}_* = \begin{bmatrix} \mathcal{H}_{W_*}^{\perp} \\ 0 \\ 0 \end{bmatrix}, \qquad \mathcal{I}_{U,W_*}\Omega\mathcal{H} = \begin{bmatrix} 0 \\ \mathcal{H} \\ 0 \end{bmatrix}, \qquad \mathcal{I}_{U,W_*}\Omega\mathcal{G} = \begin{bmatrix} 0 \\ 0 \\ L^2(\mathcal{F}_d, \mathcal{E}) \end{bmatrix}.$$

As $\mathcal{K} = \mathcal{G}_* \oplus \mathcal{H} \oplus \mathcal{G}$, we therefore have

$$\mathcal{I}_{U,W_*} \operatorname{im}\Omega = \begin{bmatrix} \mathcal{H}_{W_*}^{\perp} \\ \mathcal{H} \\ L^2(\mathcal{F}_d, \mathcal{E}) \end{bmatrix} = \mathcal{I}_{U,W_*}\mathcal{T}_{U,W_*}.$$

By the injectivity of $\mathcal{I}_{U,W_*}$, we see that the containment (4.2.11) is actually an equality:

$$\operatorname{im}\Omega = \mathcal{T}_{U,W_*}.$$

For $k \in \mathcal{G}$ and $(u, x, y) = \Omega(k)$, on the one hand we know that

$$\widehat{u} = \Phi(k) = W u^{\wedge+} \in \mathcal{H}_W, \tag{4.2.12}$$

while, on the other hand, we know that

$$(u, x, y) = (\mathcal{I}_{U,W_*})^{-1} \begin{bmatrix} 0 \\ 0 \\ u^{\wedge+} \end{bmatrix}.$$

Reading off from the first formula in (4.2.4), we get

$$\widehat{u}(z, \zeta) = \widehat{W}'(z, \zeta) u^{\wedge+}(z). \tag{4.2.13}$$

Comparing (4.2.12) with (4.2.13) and using the arbitrariness of $u^{\wedge+} \in L^2(\mathcal{F}_d, \mathcal{E})$ then leaves us with $W' = W$ as asserted.

We compute the scattering function $S \colon \mathcal{L}_W \to \mathcal{L}_{W_*}$ as follows. We first specialize to the case $k \in \mathcal{G}$ with $(u, x, y) = \Omega(k)$. From the definitions we have

$$S \colon W u^{\wedge+} \mapsto W_* y^{\wedge+}$$

where necessarily $y^{\wedge+}(z) = T(z) u^{\wedge+}(z)$. Since $S \colon \mathcal{L}_W \to \mathcal{L}_{W_*}$ intertwines the model row unitary d-tuple $\mathcal{U}_W$ on $\mathcal{L}_W$ with the model row-unitary d-tuple $\mathcal{U}_{W_*}$ on

$\mathcal{L}_{W_*}$ (and similarly, $\mathcal{U}_W^*$ with $\mathcal{U}_{W_*}^*$) by construction, necessarily, in the notation of Section 2.4, we have $S = L_T^{W,W_*}$. Thus for $p \in \mathcal{P}(\mathcal{F}_d \times \mathcal{F}_d, \mathcal{E})$, we have

$$S\colon W[p] \mapsto W_* L_T[p] := L_T^{W,W_*}[p]$$

as asserted, and Proposition 4.2.2 follows. □

We next identify an explicit scattering system $\mathfrak{S} = \mathfrak{S}_{U,W_*}$ having a preassigned pair (U, W_*) for its associated Cuntz unitary colligation $U = U(\mathfrak{S})$ (as in (3.3.2)) and incoming Cuntz shift weight W_*. To this end we define

$$\mathfrak{S}_{U,W_*} = (\boldsymbol{\mathcal{U}}_{U,W_*} = (\boldsymbol{\mathcal{U}}_{U,W_*;1}, \cdots, \boldsymbol{\mathcal{U}}_{U,W_*;d}); \boldsymbol{\mathcal{K}}_{U,W_*}, \boldsymbol{\mathcal{G}}_{U,W_*}, \boldsymbol{\mathcal{G}}_{U,W_*;*}) \tag{4.2.14}$$

where

$$\boldsymbol{\mathcal{K}}_{U,W_*} = \begin{bmatrix} \mathcal{H}_{W_*}^{\perp} \\ \mathcal{H} \\ L^2(\mathcal{F}_d, \mathcal{E}) \end{bmatrix}, \quad \boldsymbol{\mathcal{G}}_{U,W_*} = \begin{bmatrix} 0 \\ 0 \\ L^2(\mathcal{F}_d, \mathcal{E}) \end{bmatrix}, \quad \boldsymbol{\mathcal{G}}_{U,W_*;*} = \begin{bmatrix} \mathcal{H}_{W_*}^{\perp} \\ 0 \\ 0 \end{bmatrix}$$

and where the d-tuple of operators $\boldsymbol{\mathcal{U}}_{U,W_*} = (\boldsymbol{\mathcal{U}}_{U,W_*;1}, \cdots, \boldsymbol{\mathcal{U}}_{U,W_*;d})$ on $\boldsymbol{\mathcal{K}}_{U,W_*}$ is given by

$$\boldsymbol{\mathcal{U}}_{U,W_*;j}\colon \begin{bmatrix} f \\ h \\ g \end{bmatrix} \mapsto \begin{bmatrix} f' \\ h' \\ g' \end{bmatrix} \tag{4.2.15}$$

where

$$\begin{aligned} f' &= P_{\mathcal{H}_{W_*}^{\perp}} \mathcal{U}_{W_*,j} f \in \mathcal{H}_{W_*}^{\perp}, \\ h' &= A_j^* h + C^*(W_*^{-1} P_{\mathcal{H}_{W_*}} \mathcal{U}_{W_*,j} f) \in \mathcal{H}, \\ g'(z) &= (B_j^* h + D^*(W_*^{-1} P_{\mathcal{H}_{W_*}} \mathcal{U}_{W_*,j} f)) z^{\emptyset} \zeta^{\emptyset} + g(z) z_j \in L^2(\mathcal{F}_d, \mathcal{E}). \end{aligned}$$

for $j = 1, \dots, d$. It turns out that the adjoint operators $\boldsymbol{\mathcal{U}}_{U,W_*}^*$ are given by

$$\boldsymbol{\mathcal{U}}_{U,W_*;j}^*\colon \begin{bmatrix} f \\ h \\ g \end{bmatrix} \mapsto \begin{bmatrix} f'' \\ h'' \\ g'' \end{bmatrix} \tag{4.2.16}$$

where

$$\begin{aligned} f'' &= \mathcal{U}_{W_*,j}^* f + W_*([Ch + Dg_{\emptyset}]\zeta_j) \in \mathcal{H}_{W_*}^{\perp}, \\ h'' &= A_j h + B_j g_{\emptyset} \in \mathcal{H}, \\ g''(z) &= g(z) z_j^{-1} \in L^2(\mathcal{F}_d, \mathcal{E}). \end{aligned}$$

for each $j = 1, \dots, d$. We then have the following result.

THEOREM 4.2.3. *For a given pair (U, W_*) with $U\colon \begin{bmatrix} \mathcal{H} \\ \mathcal{E} \end{bmatrix} \to \begin{bmatrix} \oplus_{j=1}^d \mathcal{H} \\ \mathcal{E}_* \end{bmatrix}$ equal to a Cuntz unitary colligation (4.1.1) and W_* equal to a Cuntz shift weight with matrix entries acting on $\mathcal{E}_*$ as above, let $\mathfrak{S}_{U,W_*}$ be the collection of operators and spaces as in (4.2.14). Then:*

(1) $\mathfrak{S}_{U,W_*}$ *is a Cuntz scattering system, with outgoing wandering subspace* $\boldsymbol{\mathcal{E}}_{U,W_*}$*, with incoming wandering subspace* $\boldsymbol{\mathcal{E}}_{U,W_*;*}$ *and with scattering subspace* $\boldsymbol{\mathcal{H}}_{U,W_*}$ *given by*

$$\boldsymbol{\mathcal{E}}_{U,W_*} = \left\{ \begin{bmatrix} 0 \\ 0 \\ ez^{\emptyset} \end{bmatrix} : e \in \mathcal{E} \right\}, \tag{4.2.17}$$

$$\boldsymbol{\mathcal{E}}_{U,W_*;*} = \left\{ \begin{bmatrix} 0 \\ h \\ ez^{\emptyset} \end{bmatrix} : A_j h + B_j e = 0 \text{ for } j = 1, \dots, d. \right\} \tag{4.2.18}$$

$$\boldsymbol{\mathcal{H}}_{U,W_*} = \begin{bmatrix} 0 \\ \mathcal{H} \\ 0 \end{bmatrix}. \tag{4.2.19}$$

Moreover, if we define unitary identification maps

$$\boldsymbol{\iota} \colon \boldsymbol{\mathcal{E}}_{U,W_*} \to \mathcal{E}, \qquad \boldsymbol{\iota}_{\mathcal{H}} \colon \boldsymbol{\mathcal{H}}_{U,W_*} \to \mathcal{H}, \qquad \boldsymbol{\iota}_* \colon \boldsymbol{\mathcal{E}}_{U,W_*;*} \to \mathcal{E}_*$$

by

$$\boldsymbol{\iota} \colon \begin{bmatrix} 0 \\ 0 \\ ez^{\emptyset} \end{bmatrix} \mapsto e, \qquad \boldsymbol{\iota}_{\mathcal{H}} \colon \begin{bmatrix} 0 \\ h \\ 0 \end{bmatrix} \mapsto h, \qquad \boldsymbol{\iota}_* \colon \begin{bmatrix} 0 \\ h \\ ez^{\emptyset} \end{bmatrix} \mapsto e_* := Ch + De \tag{4.2.20}$$

then the adjusted unitary colligation associated with $\mathfrak{S}_{U,W_*}$

$$U' = \begin{bmatrix} \oplus_{j=1}^d \boldsymbol{\iota}_{\mathcal{H}} & 0 \\ 0 & \boldsymbol{\iota}_* \end{bmatrix} U(\mathfrak{S}_{U,W_8}) \begin{bmatrix} \boldsymbol{\iota}_{\mathcal{H}}^* & 0 \\ 0 & \boldsymbol{\iota}^* \end{bmatrix} := \begin{bmatrix} A_1' & B_1' \\ \vdots & \vdots \\ A_d' & B_d' \\ C' & D' \end{bmatrix} : \begin{bmatrix} \mathcal{H} \\ \mathcal{E} \end{bmatrix} \to \begin{bmatrix} \oplus_{j=1}^d \mathcal{H} \\ \mathcal{E}_* \end{bmatrix} \tag{4.2.21}$$

defined by

$$\begin{bmatrix} A_j' & B_j' \end{bmatrix} = \boldsymbol{\iota}_{\mathcal{H}} P_{\mathcal{H}} \boldsymbol{\mathcal{U}}_{U,W_*,j}^* \begin{bmatrix} I_{\mathcal{H}} & \boldsymbol{\iota}^* \end{bmatrix},$$
$$\begin{bmatrix} C' & D' \end{bmatrix} = \boldsymbol{\iota}_* P_{\mathcal{E}_*} \begin{bmatrix} I_{\mathcal{H}} & \boldsymbol{\iota}^* \end{bmatrix}$$

is identical to U*, and the adjusted incoming Cuntz shift weight* W_*' *for* $\mathfrak{S}_{U,W_*}$ *given by*

$$W'_{*v,w;\alpha,\beta} = \boldsymbol{\iota}_* P_{\boldsymbol{\mathcal{E}}_{U,W_*;*}} \boldsymbol{\mathcal{U}}_{U,W_*}^{w} \boldsymbol{\mathcal{U}}_{U,W_*}^{*v} \boldsymbol{\mathcal{U}}_{U,W_*}^{\alpha^{\top}} \boldsymbol{\mathcal{U}}_{U,W_*}^{*\beta^{\top}} \boldsymbol{\iota}_*^* \colon \mathcal{E}_* \to \mathcal{E}_* \tag{4.2.22}$$

is identical to W_*.

(2) *If* U *is the Cuntz unitary colligation* $U = U(\mathfrak{S})$ *and* W_* *is the incoming Cuntz shift weight associated with a Cuntz scattering system*

$$\mathfrak{S} = (\mathcal{U} = (\mathcal{U}_1, \dots, \mathcal{U}_d); \mathcal{K}, \mathcal{G}, \mathcal{G}_*),$$

then the map $\boldsymbol{\mathcal{I}}_{\mathfrak{S}} \colon \mathcal{K} \to \boldsymbol{\mathcal{K}}_{U,W_*}$ *given by*

$$\boldsymbol{\mathcal{I}}_{\mathfrak{S}} \colon k \mapsto \begin{bmatrix} \Phi_* P_{\mathcal{G}_*} k \\ P_{\mathcal{H}} k \\ \sum_{v \in \mathcal{F}_d} P_{\mathcal{E}}(\mathcal{U}^{*v} k) z^v \end{bmatrix} \tag{4.2.23}$$

establishes a unitary equivalence between $\mathfrak{S}$ *and* $\mathfrak{S}_{U,W_*}$.

REMARK 4.2.4. As a consequence of part (2) of Proposition 4.2.2, we conclude that, in the setting of Theorem 4.2.3, the adjusted outgoing Cuntz weight

$$W'_{v,w;\alpha,\beta} = \iota P_{\mathcal{E}_{U,W_*}} \mathcal{U}^{w}_{U,W_*} \mathcal{U}^{*v}_{U,W_*} \iota^*$$

has symbol $\widehat{W}'(z,\zeta)$ given by (4.2.5)

$$\begin{aligned} W'(z,\zeta) = {} & B^* Z_r(\zeta)^*(\zeta)^*(I - A^* Z_r(\zeta)^*)^{-1}(I - Z_r(z)A)^{-1} Z_r(z) B \\ & + T(\zeta)(\widehat{W}_*(z,\zeta) - I)T(z) + I \end{aligned}$$

with adjusted scattering function $S' \colon \mathcal{L}_{W'} \to \mathcal{L}_{W'_*}$ given by

$$S' = L_T^{W',W'_*} \colon \mathcal{L}_{W'} \to \mathcal{L}_{W'_*},$$

or equivalently (by (2.4.10)),

$$S' \colon W'[p] \mapsto V[p] \text{ for } p \in \mathcal{P}(\mathcal{F}_d \times \mathcal{F}_d)$$

where V is the $[*]$-Haplitz operator with symbol $\widehat{V}(z,\zeta) = \widehat{W}'_*(z,\zeta)T(z)$ (with $T(z)$ equal to the characteristic function $T_{\Sigma(U)}(z)$ of U).

PROOF. That $\mathcal{U}_{U,W_*}$ given by (4.2.15) is row-unitary with adjoint $\mathcal{U}^*_{U,W_*}$ given by (4.2.16) is a direct verification. It is clear that $\mathcal{U}_{U,W_*}|_{\mathcal{G}_{U,W_*}} \cong S^R|_{L^2(\mathcal{F}_d,\mathcal{E})}$ is a row shift with wandering subspace

$$\mathcal{E}_{U,W_*} := \mathcal{G}_{U,W_*} \ominus \begin{bmatrix} \mathcal{U}_{U,W_*;1} & \cdots & \mathcal{U}_{U,W_*;d} \end{bmatrix} (\oplus_{j=1}^d \mathcal{G}_{U,W_*}) = \begin{bmatrix} 0 \\ 0 \\ \mathcal{E} z^{\emptyset} \end{bmatrix}$$

as asserted in (4.2.17). From the formula for $\mathcal{U}^*_{U,W_*}$ we see that $\mathcal{G}_{U,W_*}$ is invariant for $\mathcal{U}^*$, and clearly $\mathcal{G}_{U,W_{*;*}}$ is orthogonal to $\mathcal{G}_{U,W_*}$ in $\mathcal{K}_{U,W_*}$ with $\mathcal{H}_{U,W_*} := \mathcal{K}_{U,W_*} \ominus [\mathcal{G}_{U,W_*;*} \oplus \mathcal{G}_{U,W_*}]$ given by (4.2.19). Clearly the maps $\iota \colon \mathcal{E}_{U,W_*} \to \mathcal{E}$ and $\iota_{\mathcal{H}} \colon \mathcal{H}_{U,W_*}$ are unitary. Again from the formula (4.2.16) for $\mathcal{U}_{U,W_*}$, we see that we recover A and B as

$$\begin{bmatrix} A_j & B_j \end{bmatrix} = \iota_{\mathcal{H}} P_{\mathcal{H}_{U,W_*}} \mathcal{U}_{U,W_*,j} \begin{bmatrix} \iota^*_{\mathcal{H}} & \iota^* \end{bmatrix}.$$

From (3.3.10) applied to the situation here, we arrive at

$$\begin{aligned} \mathcal{E}_{U,W_*;*} &:= \begin{bmatrix} \mathcal{U}_{U,W_*;1} & \cdots & \mathcal{U}_{U,W_*;d} \end{bmatrix} (\oplus_{j=1}^d \mathcal{G}_{U,W_*;*}) \ominus \mathcal{G}_{U,W_*;*} \\ &= [\mathcal{H}_{U,W_*} \oplus \mathcal{E}_{U,W_*}] \ominus \begin{bmatrix} \mathcal{U}_{U,W_*;1} & \ldots \mathcal{U}_{U,W_*;d} \end{bmatrix} (\oplus_{j=1}^d \mathcal{H}_{U,W_*}) \\ &= \left\{ \begin{bmatrix} 0 \\ h \\ e \end{bmatrix} : A_j h + B_j e = 0 \text{ for } j = 1, \ldots, d \right\} \end{aligned}$$

and the formula (4.2.18) for $\mathcal{E}_{U,W_*;*}$ follows. From the unitary property of U, we see that $\|Ch + De\|^2 = \|h\|^2 + \|e\|^2$ if and only if $A_j h + B_j e = 0$ for $j = 1, \ldots, d$ and that $\begin{bmatrix} C & D \end{bmatrix}$ maps such vectors $h \oplus e$ surjectively only $\mathcal{E}_*$; from this observation we see that the map $\iota_* \colon \mathcal{E}_{U,W_*;*} \to \mathcal{E}_*$ in (4.2.20) is unitary as asserted. From this characterization (4.2.18) of the space $\mathcal{E}_{U,W_*;*}$, we see that, for $h \in \mathcal{H}$ and $e \in \mathcal{E}$,

$$\begin{aligned} \iota_* P_{\mathcal{E}_{U,W_*;*}} \begin{bmatrix} \iota^*_{\mathcal{H}} & \iota^* \end{bmatrix} \begin{bmatrix} h \\ e \end{bmatrix} &= \begin{bmatrix} C & D \end{bmatrix} P_{(\ker \begin{bmatrix} C & D \end{bmatrix})^{\perp}} \begin{bmatrix} h \\ e \end{bmatrix} \\ &= \begin{bmatrix} C & D \end{bmatrix} \begin{bmatrix} h \\ e \end{bmatrix}. \end{aligned}$$

We have now completed the verification of the identity (4.2.21).

From the formula (4.2.16) for $\boldsymbol{\mathcal{U}}_{U,W_*;j}$ we see that

$$\boldsymbol{\mathcal{U}}^*_{U,W_*;j}\boldsymbol{\mathcal{E}}_{U,W_*} = \begin{bmatrix} W_*(\mathcal{E}_*\zeta_j) \\ 0 \\ 0 \end{bmatrix} \quad \text{for } j = 1, \dots, d.$$

One can easily verify that the model space $\mathcal{L}_{W_*}$ has the property

$$\underset{j=1,\dots,d;v,w\in\mathcal{F}_d}{\text{cl. span}} \quad \mathcal{U}^w_{W_*}\mathcal{U}^{*v}_{W_*}(W_*\mathcal{E}_*\zeta_j) = \mathcal{H}^\perp_{W_*}.$$

From this property it follows that $\boldsymbol{\mathcal{G}}_{U,W_*;*} \subset cl.span_{v,w\in\mathcal{F}_d}\, \mathcal{U}^w\mathcal{U}^{*v}\boldsymbol{\mathcal{E}}_{U,W_*;*}$ and we have completed the verification that $\mathfrak{S}_{U,W_*}$ given by (4.2.14) is indeed a Cuntz scattering system.

From the formula (4.2.16) for $\boldsymbol{\mathcal{U}}_{U,W_*;j}$ on $\boldsymbol{\mathcal{K}}_{U,W_*}$ and the formula (2.2.14) for $\mathcal{U}^*_{W_*,j}$ on $\mathcal{L}_{W_*}$, we see that, for $e_* \in \mathcal{E}_*$,

$$\boldsymbol{\mathcal{U}}^*_{U,W_*;j}\iota^*_*e_* = \begin{bmatrix} W_*(e_*\zeta_j) \\ 0 \\ 0 \end{bmatrix} = \begin{bmatrix} \mathcal{U}^*_{W_*;j}(W_*e_*) \\ 0 \\ 0 \end{bmatrix} \quad \text{for } j = 1, \dots, d$$

and hence the map

$$\mathcal{I}_{\widetilde{\mathcal{G}}_*} : \begin{bmatrix} f \\ 0 \\ 0 \end{bmatrix} + \sum_{v\in\mathcal{F}_d} \boldsymbol{\mathcal{U}}^v_{U,W_*}\iota^*_*e_{*v} \mapsto f(z,\zeta) + W_*\left(\sum_{v\in\mathcal{F}_d} e_{*v}z^v\right)$$

(where $f \in \mathcal{H}^\perp_{W_*}$ and $e_{*v} \in \mathcal{E}_*$ with $\sum_{v\in\mathcal{F}_d}\|e_{*v}\|^2 < \infty$) is unitary from

$$\widetilde{\boldsymbol{\mathcal{G}}}_{U,W_*;*} := \underset{v,w\in\mathcal{F}_d}{\text{cl. span}} \mathcal{U}^w_{U,W_*}\mathcal{U}^v_{U,W_*}\boldsymbol{\mathcal{E}}_{U,W_*;*}$$

onto $\mathcal{L}_{W_*}$. It is also easy to verify the intertwining

$$\mathcal{I}_{\widetilde{\mathcal{G}}_*}\left(\boldsymbol{\mathcal{U}}_{U,W_*;j}|_{\widetilde{\boldsymbol{\mathcal{G}}}_{U,W_*;*}}\right) = \mathcal{U}_{W_*;j}\mathcal{I}_{\widetilde{\mathcal{G}}_*} \text{ for } j = 1, \dots, d.$$

From this identification the identity of W'_* in (4.2.22) with W_* now follows easily.

It remains only to verify part (2) in the statement of Theorem 4.2.3. We therefore assume that we are given a Cuntz scattering system $\mathfrak{S}$ as in (3.1.1). From the formula (3.3.38) for $\|k\|^2$ in Theorem 3.3.8 we see that the map $\mathcal{I}_{\mathfrak{S}}$ given by (4.2.23) is isometric. Since $\mathcal{U}|_{\mathcal{G}}$ is a row shift by hypothesis, it is clear that $\mathcal{I}_{\mathfrak{S}}$ maps $\mathcal{G}$ onto $\boldsymbol{\mathcal{G}}_{U,W_*}$. By definition $\mathcal{I}_{\mathfrak{S}}$ maps $\mathcal{H}$ onto $\boldsymbol{\mathcal{H}}_{U,W_*}$, and by the converse side of Theorem 3.3.7 we know that $\mathcal{I}_{\mathfrak{S}}$ maps $\mathcal{G}_*$ onto $\boldsymbol{\mathcal{G}}_{U,W_*;*}$. Hence $\mathcal{I}_{\mathfrak{S}} : \mathcal{K} \to \boldsymbol{\mathcal{K}}_{U,W_*}$ is unitary. It remains only to check the intertwining relations

$$\mathcal{I}_{\mathfrak{S}}\mathcal{U}_jk = \boldsymbol{\mathcal{U}}_{U,W_*;j}\mathcal{I}_{\mathfrak{S}}k, \qquad \mathcal{I}_{\mathfrak{S}}\mathcal{U}^*_jk = \boldsymbol{\mathcal{U}}^*_{U,W_*;j}\mathcal{I}_{\mathfrak{S}}k \text{ for } j = 1, \dots, d \text{ and for } k \in \mathcal{K}.$$

It is a relatively straightforward exercise (which we leave to the reader) to check these relations for each of the three separate cases $k \in \mathcal{G}$, $k \in \mathcal{H}$ and $k \in \mathcal{G}_*$. This completes the proof of Theorem 4.2.3. □

REMARK 4.2.5. For the classical $d = 1$ case, the matrix representations (4.2.15) and (4.2.16) for $\boldsymbol{\mathcal{U}} := \boldsymbol{\mathcal{U}}_1$ and $\boldsymbol{\mathcal{U}}^*$ amount to the Schäffer matrix construction for the minimal unitary dilation of a contraction operator $\mathbf{T} := A^*$ on the Hilbert space $\mathcal{H}$, where one takes the unitary colligation U to have the form of the Halmos unitary dilation of $\mathbf{T}^*$:

$$U = \begin{bmatrix} \mathbf{T}^* & D_{\mathbf{T}} \\ D_{\mathbf{T}^*} & -\mathbf{T} \end{bmatrix} : \begin{bmatrix} \mathcal{H} \\ \mathcal{D}_{\mathbf{T}} \end{bmatrix} \to \begin{bmatrix} \mathcal{H} \\ \mathcal{D}_{\mathbf{T}^*} \end{bmatrix}.$$

Here we have set $D_{\mathbf{T}}$ equal to the *defect operator* for $\mathbf{T}$ ($D_{\mathbf{T}} = (I - \mathbf{T}^*\mathbf{T})^{1/2}$) and $\mathcal{D}_{\mathbf{T}}$ equal to the closure of the range of $D_{\mathbf{T}}$, with analogous conventions for $D_{\mathbf{T}^*}$ and $\mathcal{D}_{\mathbf{T}^*}$. Given a row contraction $\mathbf{T} = (\mathbf{T}_1, \dots, \mathbf{T}_d)$, one can again form the Halmos unitary dilation of $A = \mathbf{T}^*$ where now we set $\mathbf{T}$ equal to the operator

$$\mathbf{T} = \begin{bmatrix} \mathbf{T}_1 & \dots & \mathbf{T}_d \end{bmatrix} : \oplus_{j=1}^d \mathcal{H} \to \mathcal{H},$$

to get a Cuntz unitary colligation

$$U_A \colon \begin{bmatrix} \mathcal{H} \\ \mathcal{D}_{\mathbf{T}} \end{bmatrix} \to \begin{bmatrix} \oplus_{j=1}^d \mathcal{H} \\ \mathcal{D}_{\mathbf{T}^*} \end{bmatrix}.$$

The compression of $\mathcal{U}_{U,W_*}$ to the bottom two components $\mathcal{H} \oplus L^2(\mathcal{F}_d, \mathcal{D}_{\mathbf{T}})$ of $\boldsymbol{\mathcal{K}}_{U,W_*}$ (so the role of W_* is eliminated) amounts to the construction of the minimal row-isometric dilation of $\mathbf{T}$ in the work of Popescu [**Po89c**]. The observation here is that the construction of a minimal row-unitary dilation $\boldsymbol{\mathcal{U}}_{U,W_*}$ is not unique, but is parametrized by the choice of shift Cuntz weight W_* on $\mathcal{E}_*$. We shall return to this topic in Chapter 5.

The Cuntz scattering system $\mathfrak{S}_{U,W_*}$ (4.2.14) can be given a more coordinate-free form at the level of trajectories as follows. Set

$$\mathfrak{S}_{U,W_*} = (\mathcal{U}_{U,W_*} = (\mathcal{U}_{U,W_*;1}, \dots, \mathcal{U}_{U,W_*;d}); \mathcal{T}_{U,W_*}, \mathcal{G}_{U,W_*}, \mathcal{G}_{U,W_*;*}) \tag{4.2.24}$$

where the space $\mathcal{T}_{U,W_*}$ is the space of admissible trajectories of U (with respect to W_*), i.e., solutions (u, x, y) of (4.1.2)–(4.1.3) on $\mathcal{F}_d \times \mathcal{F}_d$ for which $u|_{\mathcal{F}_d \times \{\emptyset\}} \in \ell^2(\mathcal{F}_d, \mathcal{E})$ and $y \in \ell_{W_*}$,

$$\begin{aligned} \mathcal{G}_{U,W_*} &= \{(u,x,y) \in \mathcal{T}_{U,W_*} : h \in h_{W_*}, x(\emptyset,\emptyset) = 0\} \\ &= (\mathcal{I}_{U,W_*})^{-1} \boldsymbol{\mathcal{G}}_{U,W_*} \end{aligned} \tag{4.2.25}$$

$$\begin{aligned} \mathcal{G}_{U,W_*;*} &= \{(u,x,y) \in \mathcal{T}_{U,W_*} : x(\emptyset,\emptyset) = 0, u|_{\mathcal{F}_d \times \{\emptyset\}} = 0\} \\ &= (\mathcal{I}_{U,W_*})^{-1} \boldsymbol{\mathcal{G}}_{U,W_*;*} \end{aligned} \tag{4.2.26}$$

$$\mathcal{U}_{U,W_*;j} \colon (u,x,y) \mapsto (\widetilde{U}_j^R u, \widetilde{U}_j^R x, \widetilde{U}_j^R y) \tag{4.2.27}$$

$$\mathcal{U}^*_{U,W_*;j} \colon (u,x,y) \mapsto (\widetilde{S}_j^{R[*]} u, \widetilde{S}_j^{R[*]} x, \widetilde{S}_j^{R[*]} y). \tag{4.2.28}$$

Here $\mathcal{I}_{U,W_*} \colon \mathcal{T}_{U,W_*} \to \boldsymbol{\mathcal{K}}_{U,W_*}$ is the map given by (4.2.3) and $\widetilde{U}_j^R, \widetilde{S}_j^{R[*]}$ are the time-domain shift operators given by (3.3.20) and (3.3.21).

THEOREM 4.2.6. *Given a Cuntz unitary colligation U as in* (4.1.1) *and a Cuntz shift weight W_* with matrix entries equal to operators on the output space $\mathcal{E}_*$ of U, let $\mathfrak{S}_{U,W_*}$ be the collection of operators and spaces as in* (4.2.24). *Then:*

(1) $\mathfrak{S}_{U,W_*}$ *is a Cuntz scattering system with outgoing wandering subspace* $\mathcal{E}_{U,W_*}$*, with incoming wandering subspace* $\mathcal{E}_{U,W_*;*}$ *and with scattering subspace* $\mathcal{H}_{U,W_*}$ *given by*

$$\begin{aligned}
\mathcal{E}_{U,W_*} &= \{(u,x,y)\colon y \in h_{W_*}, x(\emptyset,\emptyset) = 0, u|_{(\mathcal{F}_d\setminus\{\emptyset\})\times\{\emptyset\}} = 0\} \\
&= (\mathcal{I}_{U,W_*})^{-1}\boldsymbol{\mathcal{E}}_{U,W_*}, \\
\mathcal{E}_{U,W_*;*} &= \{(u,x,y)\colon y \in h_{W_*}, u|_{(\mathcal{F}_d\setminus\{\emptyset\})\times\{\emptyset\}} = 0, \\
&\qquad \|(u,x,y)\|^2_{\mathcal{I}_{U,W_*}} = \|x(\emptyset,\emptyset)\|^2 + \|u(\emptyset,\emptyset)\|^2\} \\
&= (\mathcal{I}_{U,W_*})^{-1}\boldsymbol{\mathcal{E}}_{U,W_*;*}, \\
\mathcal{H}_{U,W_*} &= \{(u,x,y)\colon y \in h_{W_*}, u \in h_W^{\perp}\} \\
&= (\mathcal{I}_{U,W_*})^{-1}\boldsymbol{\mathcal{H}}_{U,W_*}.
\end{aligned}$$

Moreover, if we define unitary identification maps

$$\iota\colon \mathcal{E}_{U,W_*} \to \mathcal{E}, \qquad \iota_{\mathcal{H}}\colon \mathcal{H}_{U,W_*} \to \mathcal{H}, \qquad \iota_*\colon \mathcal{E}_{U,W_*;*} \to \mathcal{E}_*$$

by

$$\begin{aligned}
\iota\colon (u,x,y) &\mapsto e \text{ if } \mathcal{I}_{U,W_*}(u,x,y) = \begin{bmatrix} 0 \\ 0 \\ ez^{\emptyset} \end{bmatrix}, \\
i_*\colon (u,x,y) &\mapsto e_* \text{ if } \mathcal{I}_{U,W_*}(u,x,y) = \begin{bmatrix} 0 \\ h \\ \widehat{u}_+^+(z) \end{bmatrix} \\
&\text{and } C(I-zA)^{-1}h + T(z)u_+^{\wedge +}(z) = e_* z^{\emptyset}, \\
i_{\mathcal{H}}\colon (u,x,y) &\mapsto h \text{ if } \mathcal{I}_{U,W_*}(u,x,y) = \begin{bmatrix} 0 \\ h \\ 0 \end{bmatrix}
\end{aligned}$$

then the adjusted unitary colligation associated with $\mathfrak{S}_{U,W_*}$

(4.2.29)

$$U' = \begin{bmatrix} \oplus_{j=1}^d \iota_{\mathcal{H}} & 0 \\ 0 & \boldsymbol{\iota}_* \end{bmatrix} U(\mathfrak{S}_{U,W_*}) \begin{bmatrix} \iota_{\mathcal{H}}^* & 0 \\ 0 & \iota^* \end{bmatrix} := \begin{bmatrix} A_1' & B_1' \\ \vdots & \vdots \\ A_d' & B_d' \\ C' & D' \end{bmatrix} \cdot \begin{bmatrix} \mathcal{H} \\ \mathcal{E} \end{bmatrix} \to \begin{bmatrix} \oplus_{j=1}^d \mathcal{H} \\ \mathcal{E}_* \end{bmatrix}$$

is identical to U*, and the adjusted incoming Cuntz shift weight* W_*' *for* $\mathfrak{S}_{U,W_*}$

$$W'_{*v,w;\alpha,\beta} = \boldsymbol{\iota}_* P_{\boldsymbol{\mathcal{E}}_{U,W_*;*}} \boldsymbol{\mathcal{U}}^w_{U,W_*} \boldsymbol{\mathcal{U}}^{*v}_{U,W_*} \boldsymbol{\mathcal{U}}^{\alpha^\top}_{U,W_*} \boldsymbol{\mathcal{U}}^{*\beta^\top}_{U,W_*} \boldsymbol{\iota}_*^*\colon \mathcal{E}_* \to \mathcal{E}_* \tag{4.2.30}$$

is identical to W_*.

(2) *If* U *is the Cuntz unitary colligation* $U = U(\mathfrak{S})$ *and* W_* *is the incoming Cuntz shift weight associated with a Cuntz scattering system*

$$\mathfrak{S} = (\mathcal{U} = (\mathcal{U}_1, \ldots, \mathcal{U}_d); \mathcal{K}, \mathcal{G}, \mathcal{G}_*),$$

then the map $\Omega\colon \mathcal{K} \to \mathcal{K}_{U,W_*}$ *given by* (3.3.18) *establishes a unitary equivalence between* $\mathfrak{S}$ *and* $\mathfrak{S}_{U,W_*}$.

REMARK 4.2.7. Just as in Remark 4.2.4 in connection with Theorem 4.2.3, it follows that the adjusted outgoing Cuntz weight

$$W'_{v,w;\alpha,\beta} = \iota P_{\mathcal{E}_{U,W_*}} \mathcal{U}^w_{U,W_*} \mathcal{U}^{*v}_{U,W_*} \iota^*$$

for $\mathfrak{S}_{U,W_*}$ in Theorem 4.2.6 has symbol $\widehat{W}'(z,\zeta)$ given by (4.2.5)

$$\begin{aligned}\widehat{W}'(z,\zeta) = I + B^* Z_r(\zeta)^*(I - A^* Z_r(\zeta)^*)^{-1}(I - Z_r(z)A)^{-1} Z_r(z)B \\ + T(\zeta)^*(\widehat{W}_*(z,\zeta) - I)T(z)\end{aligned}$$

with adjusted scattering function $S' \colon \mathcal{L}_{W'} \to \mathcal{L}_{W'_*}$ given by

$$S' = L_T^{W',W'_*} \colon \mathcal{L}_{W'} \to \mathcal{L}_{W'_*},$$

or, equivalently,

$$S' \colon W'[p] \mapsto V[p] \text{ for } p \in \mathcal{P}(\mathcal{F}_d \times \mathcal{F}_d)$$

where V is the [*]-Haplitz operator with symbol $\widehat{V}(z,\zeta) = \widehat{W}'_*(z,\zeta)T(z)$ and where $T(z) = T_{\Sigma(U)}(z)$ is the characteristic function $T_{\Sigma(U)}(z)$ of U.

PROOF. Part (2) of the statement of the Theorem is essentially a restatement of Theorem 3.3.7. Thus it suffices to verify part (1) in the statement of Theorem 4.2.6.

All the statements of part (1) in the statement of the theorem essentially follow from the information in Theorem 4.2.3 once we verify that the map $\mathcal{I}_{U,W_*} \colon \mathcal{T}_{U,W_*} \to \boldsymbol{\mathcal{K}}_{U,W_*}$ establishes a unitary equivalence between the collection $\mathfrak{S}_{U,W_*}$ and the Cuntz scattering system $\boldsymbol{\mathfrak{S}}_{U,W_*}$ of Theorem 4.2.3. From Lemma 4.2.1 we know that $\mathcal{I}_{U,W_*}$ is unitary from $\mathcal{T}_{U,W_*}$ to $\boldsymbol{\mathcal{K}}_{U,W_*}$. Essentially by definition, $\mathcal{I}_{U,W_*}\mathcal{G}_{U,W_*} = \boldsymbol{\mathcal{G}}_{U,W_*}$, $\mathcal{I}_{U,W_*}\mathcal{G}_{U,W_*;*} = \boldsymbol{\mathcal{G}}_{U,W_*;*}$, $\mathcal{I}_{U,W_*}\mathcal{H}_{U,W_*} = \boldsymbol{\mathcal{H}}_{U,W_*}$, $\mathcal{I}_{U,W_*}\mathcal{E}_{U,W_*} = \boldsymbol{\mathcal{E}}_{U,W_*}$, $\mathcal{I}_{U,W_*}\mathcal{E}_{U,W_*;*} = \boldsymbol{\mathcal{E}}_{U,W_*;*}$. The only crucial properties remaining to be verified are the intertwining relations

$$\mathcal{I}_{U,W_*;j}\mathcal{U}_{U,W_*} = \boldsymbol{\mathcal{U}}_{U,W_*;j}\mathcal{I}_{U,W_*}, \tag{4.2.31}$$

$$\mathcal{I}_{U,W_*}\mathcal{U}^*_{U,W_*;j} = \boldsymbol{\mathcal{U}}^*_{U,W_*;j}\mathcal{I}_{U,W_*} \tag{4.2.32}$$

for $j = 1, \dots, d$.

We first verify that $(u', x', y') = \mathcal{U}_{U,W_*;j}(u, x, y)$ satisfies the system equations (4.1.2)–(4.1.3). Therefore assume that (u, x, y) satisfies (4.1.2)–(4.1.3), and set $(u', x', y') = (\widetilde{U}^R_k u, \widetilde{U}^R_k x, \widetilde{U}^R_k y)$ for some $k \in \{1, \dots, d\}$. We check first that

$$\begin{aligned} x'(g_j v, \emptyset) &= A_j x'(v, \emptyset) + B_j u'(v, \emptyset) \\ y'(v, \emptyset) &= C x'(v, \emptyset) + D u'(v, \emptyset). \end{aligned} \tag{4.2.33}$$

If $v = \emptyset$, use (3.3.20) to see that (4.2.33) becomes

$$\begin{aligned} \delta_{k,j} x(\emptyset, \emptyset) &= A_j x(\emptyset, g_k) + B_j u(\emptyset, g_k) \\ y(\emptyset, g_k) &= C x(\emptyset, g_k) + D u(\emptyset, g_k), \end{aligned}$$

or

$$\begin{bmatrix} 0 \\ \vdots \\ 0 \\ x(\emptyset,\emptyset) \\ 0 \\ \vdots \\ 0 \\ y(\emptyset, g_k) \end{bmatrix} = \begin{bmatrix} A_1 & B_1 \\ \vdots & \vdots \\ A_d & B_d \\ C & D \end{bmatrix} \begin{bmatrix} x(\emptyset, g_k) \\ u(\emptyset, g_k) \end{bmatrix}$$

where $x(\emptyset,\emptyset)$ appears in the k-th slot on the left. Since $U = \begin{bmatrix} A & B \\ C & D \end{bmatrix}$ is unitary, multiplication on the left by U^* converts this last expression into the equivalent form

$$A_k^* x(\emptyset,\emptyset) + C^* y(\emptyset, g_k) = x(\emptyset, g_k)$$
$$B_k^* x(\emptyset,\emptyset) + D^* y(\emptyset, g_k) = u(\emptyset, g_k)$$

which is just (4.1.3) with $v = \emptyset$ and $w = \emptyset$. Thus (4.2.33) holds for the case $v = \emptyset$. If $v \neq \emptyset$, say $v = v' g_\ell$ for some $v' \in \mathcal{F}_d$ and $\ell \in \{1, \dots, d\}$. Then from (3.3.20) we see that (4.2.33) becomes

$$\begin{aligned} \delta_{\ell,k} x(g_j v', \emptyset) &= \delta_{\ell,k} A_j x(v', \emptyset) + \delta_{\ell,k} B_j u(v', \emptyset) \\ \delta_{\ell,k} y(v', \emptyset) &= \delta_{\ell,k} C x(v', \emptyset) + \delta_{\ell,k} D u(v', \emptyset). \end{aligned} \tag{4.2.34}$$

If $\ell \neq k$, (4.2.34) collapses to the tautology $0 = 0$. If $\ell = k$, (4.2.34) collapses to (4.1.2) with v' in place of v. Thus (4.2.33) holds in all cases.

We next wish to verify that (u', x', y') satisfies (4.1.3), i.e.,

$$\begin{aligned} x'(v, g_j w) &= A_j^* x'(v, w) + C^* y'(v, g_j w) \\ u'(v, g_j w) &= B_j^* x'(v, w) + D^* y'(v, g_j w) \end{aligned} \tag{4.2.35}$$

If $v = \emptyset$, from (3.3.20) we see that (4.2.35) becomes

$$x(\emptyset, g_j w g_k) = A_j^* x(\emptyset, w g_k) + C^* y(\emptyset, g_j w g_k)$$
$$u(\emptyset, g_j w g_k) = B_j^* x(\emptyset, w g_k) + D^* y(\emptyset, g_j w g_k)$$

which is just (4.1.3) with $v = \emptyset$ and $w g_k$ in place of w. If $v \neq \emptyset$, assume that $v = v' g_\ell$ for some $v' \in \mathcal{F}_d$ and $\ell \in \{1, \dots, d\}$. Again use (3.3.20) to see that (4.2.35) becomes

$$\begin{aligned} \delta_{\ell,k} x(v', g_j w) &= \delta_{\ell,k} A_j^* x(v', w) + \delta_{\ell,k} C^* y(w', g_j w) \\ \delta_{\ell,k} u(v', g_j w) &= \delta_{\ell,k} B_j^* x(v', w) + \delta_{\ell,k} D^* y(v', g_j w). \end{aligned} \tag{4.2.36}$$

For $\ell \neq k$, (4.2.36) is the tautology $0 = 0$. For $\ell = k$, (4.2.36) collapses to (4.1.3) with v' in place of v. We conclude that (u', x', y') satisfies (4.1.2) and (4.1.3) in all cases. From the Haplitz structure of W_* and (2.2.18) we see that U_j^R preserves $\mathcal{L}_{W_*}$. Hence $\mathcal{T}_{U,W_*}$ is invariant under $\mathcal{U}_{U,W_*;k}$ for each $k = 1, \dots, d$ as asserted.

Temporarily define an operator $\mathcal{V}_{U,W_*;k}$ by the right hand side of formula (4.2.28)

$$\mathcal{V}_{U,W_*;k} \colon (u, x, y) \mapsto (u'', x'', y'') := (S_k^{R[*]} u, S_k^{R[*]} x, S_k^{[*]} y).$$

Assume that (u, x, y) satisfies (4.1.2) and (4.1.3); we wish to check that (u'', x'', y'') satisfies (4.1.2) and (4.1.3), i.e.,

$$\begin{aligned} x''(g_j v, \emptyset) &= A_j x''(v, \emptyset) + B_j u''(v, \emptyset) \\ y''(v, \emptyset) &= C x''(v, w) + D u''(v, \emptyset) \end{aligned} \tag{4.2.37}$$

and

$$\begin{aligned} x''(v, g_j w) &= A_j^* x''(v, w) + C^* y''(v, g_j w) \\ u''(v, g_j w) &= B_j^* x''(v, w) + D^* y''(w, g_j w). \end{aligned} \tag{4.2.38}$$

Note that (4.2.37) collapses to (4.1.2) while (4.2.38) collapses to (4.1.3) with vg_k in place of v. Moreover, from (2.2.19) we see that $\mathcal{L}_{W_*}$ is invariant under $S_k^{R[*]}$ for each $k = 1, \dots, d$. We conclude that $\mathcal{T}_{U,W_*}$ is also invariant under $\mathcal{V}_{U,W_*;k}$ for each $k = 1, \dots, d$.

We next verify the intertwining relation (4.2.31). Application of the definitions gives

$$\mathcal{I}_{U,W_*} \circ \mathcal{U}_{U,W_*;j}(u, x, y) = \begin{bmatrix} P_{\mathcal{H}_{W_*}^\perp} \left(\sum_w y(\emptyset, w g_j) \zeta^w + \sum_{v', w} y(v', w) z^{v'} z_j \zeta^w \right) \\ x(\emptyset, g_j) \\ u(\emptyset, g_j) + \sum_{v'} u(v', \emptyset) z^{v'} z_j \end{bmatrix} \tag{4.2.39}$$

where

$$\begin{aligned} x(\emptyset, g_j) &= A_j^* x(\emptyset, \emptyset) + C^* y(\emptyset, g_j) \\ u(\emptyset, g_j) &= B_j^* x(\emptyset, \emptyset) + D^* y(\emptyset, g_j) \end{aligned}$$

by (4.1.3). On the other hand, from the definitions we see that

$$\begin{aligned} &\mathcal{U}_{U,W_*;j} \mathcal{I}_{U,W_*}(u, x, y) \\ &= \begin{bmatrix} P_{\mathcal{H}_{W_*}^\perp} \mathcal{U}_{W_*,j} (P_{\mathcal{H}_{W_*}^\perp} \widehat{y}) \\ A_j^* x(\emptyset, \emptyset) + C^* \left(W_*^{-1} P_{\mathcal{H}_{W_*}} \mathcal{U}_{W_*,j} P_{\mathcal{H}_{W_*}^\perp} \widehat{y} \right) \\ B_j^* x(\emptyset, \emptyset) + D^* \left(W_*^{-1} P_{\mathcal{H}_{W_*}} \mathcal{U}_{W_*,j} P_{\mathcal{H}_{W_*}^\perp} \widehat{y} \right) + \sum_v u(v, \emptyset) z^v z_j \end{bmatrix}. \end{aligned} \tag{4.2.40}$$

Note that the bottom two components of (4.2.39) agree with the bottom two components of (4.2.40) once we verify that

$$y(\emptyset, g_j) = W_*^{-1} P_{\mathcal{H}_{W_*}} \mathcal{U}_{W_*,j} (P_{\mathcal{H}_{W_*}^\perp} \widehat{y}). \tag{4.2.41}$$

To verify (4.2.41), we compute (by using the explicit formula (2.3.10) for $P_{\mathcal{H}_{W_*}}$)

$$\begin{aligned} P_{\mathcal{H}_{W_*}^\perp} \widehat{y} &= \widehat{y} - P_{\mathcal{H}_{W_*}} \widehat{y} \\ &= \sum_{v,w} y(v, w) z^v \zeta^w - W_* \left[\sum_v y(v, \emptyset) z^v \right] \end{aligned}$$

and hence

$$\begin{aligned}
\mathcal{U}_{W_*,j}(P_{\mathcal{H}_{W_*}^\perp}\widehat{y}) &= U_j^R(P_{\mathcal{H}_{W_*}^\perp}\widehat{y}) \\
&= \sum_w y(\emptyset, wg_j)\zeta^w + \sum_{v',w} y(v',w)z^v z_j \zeta^w - W_* \left[S_j^R(\sum_v y(v,\emptyset)z^v) \right] \\
&= \sum_w y(\emptyset, wg_j)\zeta^w + \sum_{v',w} y(v',w)z^{v'} z_j \zeta^w - W_* \left[\sum_v y(v,\emptyset)z^v z_j \right]. \qquad (4.2.42)
\end{aligned}$$

Thus

$$\begin{aligned}
P_{\mathcal{H}_{W_*}}\mathcal{U}_{W_*,j}(P_{\mathcal{H}_{W_*}^\perp}\widehat{y}) &= W_* \left[y(\emptyset, g_j) + \sum_{v'} y(v',\emptyset)z^{v'} z_j - W_*^+ \left[\sum_v y(v,\emptyset)z^v z_j \right] \right] \\
&= W_*[y(\emptyset, g_j)] \qquad (4.2.43)
\end{aligned}$$

(where we used that W_* is a Haplitz extension of the identity, so $W_*^+ = I$) and (4.2.41) now follows.

To check that the top components of (4.2.39) and (4.2.40) match, we proceed as follows. From (4.2.42) and (4.2.43) we have

$$\begin{aligned}
P_{\mathcal{H}_{W_*}^\perp}\mathcal{U}_{W_*,j}(P_{\mathcal{H}_{W_*}^\perp}\widehat{y}) &= \mathcal{U}_{W_*,j}(P_{\mathcal{H}_{W_*}^\perp}\widehat{y}) - P_{\mathcal{H}_{W_*}}\mathcal{U}_{W_*,j}(P_{\mathcal{H}_{W_*}^\perp}\widehat{y}) \\
&= \sum_w y(\emptyset, wg_j)\zeta^w + \sum_{v',w} y(v',w)z^{v'} z_j \zeta^w \\
&\quad - W_* \left[\sum_v y(v,\emptyset)z^v z_j \right] - W_*\left[y(\emptyset, g_j)\right]. \qquad (4.2.44)
\end{aligned}$$

On the other hand, again by using (2.3.10) we have

$$\begin{aligned}
&P_{\mathcal{H}_{W_*}^\perp}\left(\sum_w y(\emptyset, wg_j)\zeta^w + \sum_{v',w} y(v',w)z^{v'} z_j \zeta^w \right) \\
&= \sum_w y(\emptyset, wg_j)\zeta^w + \sum_{v',w} y(v',w)z^{v'} z_j \zeta^w - W_* \left(y(\emptyset, g_j) + \sum_{v'} y(v',\emptyset)z^{v'} z_j \right). \qquad (4.2.45)
\end{aligned}$$

Comparison of (4.2.44) and (4.2.45) shows that the top components of (4.2.39) and (4.2.40) agree as well. We have now verified (4.2.31).

We next seek to verify the parallel intertwining relation (4.2.32). From the definitions we have

$$\mathcal{I}_{U,W_*} \circ \mathcal{V}_{U,W_*;,j}(u,x,y) = \begin{bmatrix} P_{\mathcal{H}_{W_*}^\perp}\left(\sum_{v,w} y(wg_j, w)z^v\zeta^w \right) \\ x(g_j,\emptyset) \\ \sum_v u(vg_j,\emptyset)z^v) \end{bmatrix} \qquad (4.2.46)$$

while

$$\boldsymbol{\mathcal{U}}_{U,W_*;j}^* \circ \mathcal{I}_{U,W_*}(u,x,y) = \begin{bmatrix} \mathcal{U}_{W_*,j}^*(P_{\mathcal{H}_{W_*}^\perp}\widehat{y}) + W_*\left([Cx(\emptyset,\emptyset) + Du(\emptyset,\emptyset)]\zeta_j \right) \\ A_j x(\emptyset,\emptyset) + B_j u(\emptyset,\emptyset) \\ \sum_v u(vg_j,\emptyset)z^v \end{bmatrix}. \qquad (4.2.47)$$

From the forward system equations (4.1.2) we know that

$$x(g_j, \emptyset) = A_j x(\emptyset, \emptyset) + B_j u(\emptyset, \emptyset).$$

Thus we see immediately that the bottom two components of (4.2.46) and (4.2.47) agree. To verify the matching of the top components, first note (again by (2.3.10)) that

$$\begin{aligned} P_{\mathcal{H}^\perp_{W_*}} \left(\sum_{v,w} y(vg_j, w) z^v \zeta^w \right) &= \sum_{v,w} y(vg_j, w) z^v \zeta^w - W_* \left[\sum_v y(vg_j, \emptyset) z^v \right] \\ &= \sum_{v,w} y(vg_j, w) z^v \zeta^w - W_* \left[U_j^{R[*]} \left[\sum_v y(v, \emptyset) z^v \right] - y(\emptyset, \emptyset) \zeta_j \right] \\ &= S_j^{R[*]} \left(\sum_{v,w} y(v, w) z^v \zeta^w - W_* \left[\sum_v y(v, \emptyset) z^v \right] \right) - W_* \left[y(\emptyset, \emptyset) \zeta_j \right] \\ &= \mathcal{U}^*_{W_*, j} (P_{\mathcal{H}^\perp_{W_*}} \widehat{y}) + W_* \left([Cx(\emptyset, \emptyset) + Du(\emptyset, \emptyset)] \zeta_j \right) \end{aligned} \tag{4.2.48}$$

where we used the second forward system equation in (4.1.2) for the last step. From (4.2.48) we see that the top components of (4.2.46) and (4.2.47) match as wanted. Consequently,

$$\mathcal{V}_{U, W_*; j} = \mathcal{I}^*_{U, W_*} \boldsymbol{\mathcal{U}}^*_{U, W_*; j} \mathcal{I}_{U, W_*} = \left(\mathcal{I}^*_{U, W_*} \boldsymbol{\mathcal{U}}_{U, W_*; j} \mathcal{I}_{U, W_*} \right)^* = \mathcal{U}^*_{U, W_*; j}$$

and we conclude that $\mathcal{V}_{U, W_*; j} = \mathcal{U}^*_{U, W_*; j}$. This completes the proof of (4.2.32) and concludes our proof of Theorem 4.2.6. □

We can now derive an alternate formula for the $\mathcal{T}_{U, W_*}$-norm.

THEOREM 4.2.8. *Let*

$$\mathfrak{S} = (\mathcal{U} = (\mathcal{U}_1, \dots, \mathcal{U}_d); \mathcal{K}, \mathcal{G}, \mathcal{G}_*)$$

be a Cuntz scattering system with associated unitary colligation U and incoming shift Cuntz weight W_*. Then for $k \in \mathcal{K}$ and $(u, x, y) = \Omega k \in \mathcal{T}_{U, W_*}$, we have

$$\|k\|^2_{\mathcal{K}} = \|\widehat{y}\|^2_{\mathcal{L}_{W_*}} + \lim_{N \to \infty} \sum_{v \colon |v| = N} \|x(v, \emptyset)\|^2_{\mathcal{H}}. \tag{4.2.49}$$

PROOF. Since $\mathcal{U}$ is a row-unitary d-tuple, we have

$$\|k\|^2_{\mathcal{K}} = \sum_{v \colon |v| = N} \|\mathcal{U}^{*v} k\|^2 \text{ for each } N = 1, 2, 3, \dots .$$

Expressing each term on the right in terms of the $\mathcal{T}_{U, W_*}$-norm gives

$$\|k\|^2 = \sum_{v \colon |v| = N} \|P_{\mathcal{H}^\perp_{W_*}} \mathcal{U}^{*v}_{W_*} \widehat{y}\|^2_{\mathcal{H}^\perp_{W_*}} + \sum_{v \colon |v| = N} \|x(v, \emptyset)\|^2_{\mathcal{H}} + \sum_{v \colon |v| > N} \|u(v, \emptyset)\|^2.$$

Letting $N \to \infty$ in this last expression leaves us with (4.2.49). □

As a corollary we have the following characterization of the space $\widetilde{\mathcal{G}}_*$.

COROLLARY 4.2.9. *With hypotheses as in Theorem 4.2.8, if $k \in \mathcal{K}$ with (u, x, y) equal to Ωk, then $k \in \widetilde{\mathcal{G}}_*$ if and only if*

$$\lim_{N \to \infty} \sum_{v \colon |v| = N} \|x(v, \emptyset)\|^2 = 0. \tag{4.2.50}$$

PROOF. We know that $k \in \widetilde{\mathcal{G}}_*$ if and only if k is in the initial space of Φ_*, i.e., if and only if $\|\widehat{y}\|^2_{\mathcal{L}_{W_*}} = \|k\|^2$. From formula (4.2.49), we see that this happens if and only if (4.2.50) happens. □

REMARK 4.2.10. In the classical case, there is an expression analogous to (4.2.49) for $\|k\|^2$ in terms of $\|u\|^2$ and limiting behavior of the state vector x as time goes to minus infinity. As a corollary one gets a characterization of the space $\widetilde{\mathcal{G}}$ in terms of the vanishing of the norm of the state vector at minus infinity. In the noncommutative setup here involving representations of the Cuntz algebra, the asymptotics at minus infinity are much murkier and there seems to be no analogous formula for $\|k\|^2$ and characterization of $\widetilde{\mathcal{G}}$ in terms of asymptotic behavior of the norm of the state at minus infinity.

We next compute symbols $\widehat{L}(z, \zeta)$ (2.4.13) and $X(z, \zeta)$ (2.4.16) associated with the admissible triple (T, W, W_*) (see Theorem 2.4.5) for the case where $T(z)$ is the characteristic function of the Cuntz unitary colligation U with W determined by U and W_* as in (4.2.5). A remarkable feature is that that $X(z, \zeta)$ and $\widehat{L}(z, \zeta)$ are completely determined by $U = \begin{bmatrix} A & B \\ C & D \end{bmatrix}$ independent of the choice of incoming Cuntz shift weight W_*; the work in Chapter 5 to come explains this phenomenon.

PROPOSITION 4.2.11. *Let U be a Cuntz unitary colligation* (4.1.1) *and W_* a shift Cuntz weight with coefficients acting on the output space $\mathcal{E}_*$ of U as above, let W be the Cuntz shift weight on the input space $\mathcal{E}$ of U with symbol $\widehat{W}(z, \zeta)$ given by* (4.2.5), *and let $T(z) = D + C(I - Z_r(z)A)^{-1}Z_r(z)B$ be the characteristic function of U. Then the $*$-defect kernel $X(z, \zeta)$ given by* (2.4.16) *and the defect kernel $\widehat{L}(z, \zeta)$ given by* (2.4.13) *for the triple (T, W, W_*) are completely determined by the colligation $U = \begin{bmatrix} A & B \\ C & D \end{bmatrix}$ independently of the choice of outgoing Cuntz weight W_* according to the formulas*

$$X(z, \zeta) = B^* Z_r(\zeta)^*(I - A^* Z_r(\zeta)^*)^{-1}(I - Z_r(z)A)^{-1} Z_r(z)B$$

$$= \sum_{k,j=1}^{d} \sum_{v,w \in \mathcal{F}_d} (B_k^* A^{*w} A^v B_j) z^v z_j \zeta_k \zeta^w \tag{4.2.51}$$

$$\widehat{L}(z, \zeta) = B^* Z_r(\zeta)^*(I - A^* Z_r(\zeta)^*)^{-1}(I - Z_r(z)A)^{-1} Z_r(z)B$$

$$+ I - T(\zeta)^* T(z) - \sum_{u \neq \emptyset} (z^{-1})^{u^\top} T(\zeta)^* T(z) (\zeta^{-1})^u \tag{4.2.52}$$

where $T(z)$ is the characteristic function (4.1.11) *of U*

$$T(z) = D + C(I - Z_r(z)A)^{-1} Z_r(z)B = D + \sum_{k=1}^{d} \sum_{v \in \mathcal{F}_d} (CA^v B_k) z^v z_k.$$

PROOF. The formula (4.2.52) for $\widehat{L}(z, \zeta)$ comes from (2.4.13) upon plugging in the formula (4.2.5) for $\widehat{W}(z, \zeta)$ and observing that the terms involving $\widehat{W}_*(z, \zeta)$ cancel. The $*$-defect kernel $X(z, \zeta)$ is then determined from the defect kernel $\widehat{L}(z, \zeta)$ according to (2.4.26), and Proposition 4.2.11 follows. □

As explained in Remark 3.2.5, the scattering function $S \in \mathcal{S}(W, W_*)$, a complete unitary invariant for a minimal scattering system $\mathfrak{S}$, can in turn be constructed from several related sets of parameters, namely: (1) (T, W, W_*), (2) (T, L, W_*) and (3) (T, X, W_*). We have now identified the $T \in \mathcal{S}_{nc,d}(\mathcal{E}, \mathcal{E}_*)$ so that $S = L_T^{W,W_*}$ as the characteristic function of the Cuntz-unitary colligation $U(\mathfrak{S})$ associated with $\mathfrak{S}$. We next present an alternate version of the Sz.-Nagy-Foiaş model $\mathfrak{S}_{NF,alt} = \mathfrak{S}_{NF,alt}(T, L, W_*)$

$$\mathfrak{S}_{NF,alt} = (\mathcal{U}_{NF,alt} = (\mathcal{U}_{NF,alt,1}, \dots, \mathcal{U}_{NF,alt,d}), \mathcal{K}_{NF'}, \mathcal{G}_{NF'}, \mathcal{G}_{NF'*}) \tag{4.2.53}$$

which is constructed from the second set of parameters (T, L, W_*). Recall that the parameter set (T, L, W_*) is admissible if $T \in \mathcal{S}_{nc,d}(\mathcal{E}, \mathcal{E}_*)$, L is a Cuntz-weight extension of the Cuntz-Toeplitz operator $I - (M_T)^* M_T$, and W_* is a Cuntz-weight extension of the identity (i.e., W_* is a shift Cuntz weight). Given such an admissible triple (T, L, W_*), define spaces

$$\mathcal{K}_{NF,alt} = \begin{bmatrix} \mathcal{L}_{W_*} \\ \mathcal{L}_L \end{bmatrix}, \qquad \mathcal{G}_{NF,alt} = \begin{bmatrix} W_* M_T \\ L \end{bmatrix} L^2(\mathcal{F}_d \times \{\emptyset\}, \mathcal{E}) \qquad \mathcal{G}_{NF,alt*} = \begin{bmatrix} \mathcal{H}_{W_*}^{\perp} \\ 0 \end{bmatrix}$$

together with row-unitary evolution operators

$$\mathcal{U}_{NF,alt,j} = \begin{bmatrix} \mathcal{U}_{W_*,j} & 0 \\ 0 & \mathcal{U}_{L,j} \end{bmatrix} \text{ for } j = 1, \dots, d.$$

Then we have the following theorem.

THEOREM 4.2.12. *Let W and W_* be Cuntz-weight extensions of the identity on auxiliary Hilbert spaces $\mathcal{E}$ and $\mathcal{E}_*$ respectively, and let $T(z)$ be a formal power series for which $S := L_T^{W,W_*}$ is in the Schur-class $\mathcal{S}(W, W_*)$. Then the system $\mathfrak{S}_{NF,alt}$ as in (4.2.53) is a minimal Cuntz scattering system with outgoing and incoming spaces given by*

$$\mathcal{E}_{NF,alt} = \begin{bmatrix} W_* L_T \\ L \end{bmatrix} \mathcal{E}, \qquad \mathcal{E}_{NF,alt*} = \begin{bmatrix} W_* \mathcal{E}_* \\ 0 \end{bmatrix}$$

with scattering function $S_{NF,alt}$ coinciding with $S = L_T^{W,W_*}$ via the identifications

$$i_{NF,alt} \colon \mathcal{E} \to \mathcal{E}_{NF,alt}, \qquad i_{NF,alt*} \colon \mathcal{E}_* \to \mathcal{E}_{NF,alt*}$$

given by

$$i_{NF,alt} \colon e \mapsto \begin{bmatrix} W_*[T(z)e] \\ L[e] \end{bmatrix}, \qquad i_{NF,alt*} \colon e_* \mapsto \begin{bmatrix} W_*[e_*] \\ 0 \end{bmatrix}.$$

Moreover, any other minimal Cuntz scattering system $\mathfrak{S}$ with scattering function equal to $S = L_T^{W,W_*}$ is unitarily equivalent to the alternate Sz.-Nagy-Foiaş-model scattering system $\mathfrak{S}_{NF,alt}$ associated with L_T^{W,W_*} via the identification map $\mathcal{I}_{NF,alt} \colon \mathcal{K} \to \mathcal{K}_{NF,alt}$ given by

$$\mathcal{I}_{NF,alt} \colon \Phi_*^* f + \Phi^* W g \mapsto \begin{bmatrix} I & W_* L_T \\ 0 & L \end{bmatrix} \begin{bmatrix} f \\ g \end{bmatrix}.$$

PROOF. Recall that the version of the Sz.-Nagy-Foiaş model in Theorem 3.2.3 has ambient space

$$\mathcal{K}_{NF} = \begin{bmatrix} \mathcal{L}_{W_*} \\ \text{clos. } D_S \mathcal{L}_W \end{bmatrix} \subset \begin{bmatrix} \mathcal{L}_{W_*} \\ \mathcal{L}_W \end{bmatrix}.$$

For an element of the form $Wg \in \mathcal{L}_W$ with $g \in L^2_{\text{fin}-}(\mathcal{F}_d \times \mathcal{F}_d, \mathcal{E})$ (a dense subset of $\mathcal{L}_W$), observe that

$$\begin{aligned}\|D_S Wg\|^2_{\mathcal{L}_W} &= \|Wg\|^2_{\mathcal{L}_W} - \|SWg\|^2_{\mathcal{L}_{W_*}} = \langle Wg, g\rangle_{L^2} - \langle W_* L_T g, L_T g\rangle_{L^2} \\ &= \langle Lg, g\rangle_{L^2} = \|Lg\|^2_{\mathcal{L}_L}\end{aligned}$$

(where we use the defining connection $L = W - L_T^{[*]} W_* L_T$ between (T, W, W_*) and (T, L, W_*)) and hence we may define a unitary map $\iota\colon \text{clos.}\, D_S \mathcal{L}_W \to \mathcal{L}_L$ by $\iota\colon D_S Wg \mapsto Lg$. Now it is a simple matter of checking that this modification of the second component of the Sz.-Nagy-Foiaş scattering system $\mathfrak{S}_{NF}$ leads to the alternate Sz.-Nagy-Foiaş scattering system $\mathfrak{S}_{NF,alt}$. $\square$

CHAPTER 5

Scattering, Systems and Dilation Theory: the Cuntz-Toeplitz Setting

5.1. Outgoing Cuntz scattering systems

In this section we analyze the structure of the forward part of a Cuntz scattering system. For this purpose we define an *outgoing Cuntz scattering system* as a collection

$$\mathfrak{S}^+ = (\mathcal{S} = (\mathcal{S}_1, \dots, \mathcal{S}_d), \mathcal{K}^+, \mathcal{G}_*^+, \mathcal{G}) \tag{5.1.1}$$

such that $\mathcal{S}$ is a row isometry on the Hilbert space $\mathcal{K}^+$, and $\mathcal{G}_*^+$ and $\mathcal{G}$ are subspaces of $\mathcal{K}^+$ such that

(1) $\mathcal{G}_*^+$ is the smallest $\mathcal{S}$-invariant subspace containing

$$\mathcal{E}_* := \mathcal{K}^+ \ominus \operatorname*{span}_{j=1,\dots,d} \mathcal{S}_j \mathcal{K}_j^+;$$

thus $S|_{\mathcal{G}_*^+}$ is a row shift and $\mathcal{G}_*^+ = \oplus_{v \in \mathcal{F}_d} \mathcal{S}^v \mathcal{E}_*$.

(2) $\mathcal{S}_{\mathcal{G}}$ is a row shift; thus $\mathcal{G} = \oplus_{v \in \mathcal{F}_d} \mathcal{S}^v \mathcal{E}$ where $\mathcal{E} = \mathcal{G} \ominus \operatorname{span}_{j=1,\dots,d} \mathcal{S}_j \mathcal{G}$.

Note that if $\mathfrak{S} = (\mathcal{U}; \mathcal{K}, \mathcal{G}_*, \mathcal{G})$ as in (3.1.1) is a Cuntz scattering system, then $\mathfrak{S}^+ = (\mathcal{S}; \mathcal{K}^+, \mathcal{G}_*^+, \mathcal{G})$ is an outgoing Cuntz scattering system, where we set

$$\begin{gathered} \mathcal{K}^+ = \mathcal{K} \ominus \mathcal{G}_*, \qquad \mathcal{G}_*^+ = \widetilde{\mathcal{G}}_* \ominus \mathcal{G}_* \text{ (where } \widetilde{\mathcal{G}}_* \text{ is given by (3.1.3))} \\ \mathcal{S} = \mathcal{U}|_{\mathcal{K}^+}. \end{gathered} \tag{5.1.2}$$

In this case, let us say that the outgoing Cuntz scattering system $\mathfrak{S}^+$ is the *forward part* of the full Cuntz scattering system $\mathfrak{S}$. Conversely, any outgoing Cuntz scattering system can be realized as the forward part of a full Cuntz scattering system by embedding the row isometry $\mathcal{S}$ on $\mathcal{H}$ into a row unitary extension $\mathcal{U}$ on a space $\mathcal{K} \supset \mathcal{H}$.

Let us say that the outgoing Cuntz scattering system is *minimal* if there is no nonzero $\mathcal{S}$-reducing subspace $\mathcal{H}_u$ inside the scattering subspace $\mathcal{H} := \mathcal{K}^+ \ominus \mathcal{G}$ on which $\mathcal{S}$ is row unitary. By Theorem 3.3.4, we see that that $\mathfrak{S}^+$ is minimal if and only if the Cuntz scattering system $\mathfrak{S}$ is minimal whenever $\mathfrak{S}^+$ arises as the forward part of $\mathfrak{S}$.

Suppose that we are given an outgoing Cuntz scattering system $\mathfrak{S}^+$ as in (5.1.1). As $\mathcal{S}$ is a row isometry, $\mathcal{S}$ has a Wold decomposition (see [**Po89c**]). By the way the Wold decomposition is constructed, we see that $\mathcal{G}_*^+$ gives the row-shift subspace, and the row-unitary subspace $\mathcal{R}$ is obtained as

$$\mathcal{R} = \mathcal{K}^+ \ominus \mathcal{G}_*^+ = \bigcap_{N=0}^{\infty} \operatorname*{span}_{v\colon |v|=N} \mathcal{S}^v \mathcal{K}^+.$$

The minimality assumption is exactly that $\mathcal{H} \cap \mathcal{R} = \{0\}$. As $\mathcal{H} = \mathcal{K}^+ \ominus \mathcal{G}$, this can be reexpressed as: *$\mathfrak{S}^+$ is minimal if and only if the subspace $P_{\mathcal{R}}\mathcal{G}$ is $*$-cyclic for the row-unitary d-tuple $\mathcal{S}|_{\mathcal{R}}$*, i.e., if and only if the linear manifold

$$\mathcal{R}_0 = \operatorname*{span}_{v,w \in \mathcal{F}_d} \mathcal{S}^w \mathcal{S}^{*v} P_{\mathcal{R}} \mathcal{G} \tag{5.1.3}$$

is dense in $\mathcal{R}$. Hence, in the minimal case, the linear manifold

$$\mathcal{D}_0 = \mathcal{G}_*^+ + \mathcal{R}_0 = \{k_1 + \mathcal{S}^w \mathcal{S}^{*v} P_{\mathcal{R}} k_2 \colon k_1 \in \mathcal{G}_*^+,\ k_2 \in \mathcal{G} \text{ and } v, w \in \mathcal{F}_d\} \tag{5.1.4}$$

is dense in $\mathcal{K}^+$.

For an outgoing Cuntz scattering system, we define only the forward part of the Fourier representations as follows. Define $\Phi^+ \colon \mathcal{K}^+ \to L^2(\mathcal{F}_d, \mathcal{E})$ and $\Phi_*^+ \colon \mathcal{K}^+ \to L^2(\mathcal{F}_d, \mathcal{E}_*)$ by

$$\Phi^+ \colon k \mapsto \sum_{v \in \mathcal{F}_d} (P_{\mathcal{E}} \mathcal{S}^{*v} k) z^v, \qquad \Phi_*^+ \colon k \mapsto \sum_{v \in \mathcal{F}_d} (P_{\mathcal{E}_*} \mathcal{S}^{*v} k) z^v.$$

Then Φ^+ is a coisometry mapping $\mathcal{K}^+$ onto $L^2(\mathcal{F}_d, \mathcal{E})$ with initial space equal to $\mathcal{G}$ while Φ_*^+ is a coisometry mapping $\mathcal{K}^+$ onto $L^2(\mathcal{F}_d, \mathcal{E}_*)$ with initial space equal to $\mathcal{G}_*^+$. Furthermore, we have the intertwining properties

$$\Phi^+ \mathcal{S}_j = S_j^R \Phi^+, \qquad \Phi_*^+ \mathcal{S}_j = S_j^R \Phi_*^+ \qquad \text{for } j = 1, \dots, d. \tag{5.1.5}$$

Given this setup, we can define the *scattering operator* as the forward part of the scattering operator for a full scattering system, namely, as $\mathbf{S}^+ \colon \mathcal{G} \to \mathcal{G}_*^+$ given by

$$\mathbf{S}^+ = P_{\mathcal{G}_*^+}|_{\mathcal{G}}.$$

As $\mathcal{G}$ and $\mathcal{G}_*^+$ are $\mathcal{S}$-invariant, it is easy to deduce the intertwining

$$\mathbf{S}^+(\mathcal{S}_j|_{\mathcal{G}}) = (\mathcal{S}_j|_{\mathcal{G}_*^+})\mathbf{S}^+ \text{ for } j = 1, \dots, d.$$

The concrete scattering operator then is given by

$$M = \Phi_*^+ \Phi^{+*} = \Phi_*^+ \mathbf{S}^+ \Phi^{+*} \colon L^2(\mathcal{F}_d, \mathcal{E}) \to L^2(\mathcal{F}_d, \mathcal{E}_*).$$

From the intertwining properties of Φ_*^+ and Φ^+ we deduce that

$$M S_j^R = S_j^R M \text{ for } j = 1, \dots, d$$

and hence M is a multiplication operator $M = M_T$ for some multiplier $T(z) = \sum_{v \in \mathcal{F}_d} T_v z^v$ which we call the *characteristic function* of $\mathfrak{S}^+$ (to avoid confusion with the scattering function $S \in \mathcal{S}(W, W_*)$ for the full scattering system $\mathfrak{S}$). It is also easily checked that $\|M_T\| = \|\mathbf{S}^+\| \leq 1$, so $T \in \mathcal{S}_{nc,d}(\mathcal{E}, \mathcal{E}_*)$. Clearly, the characteristic function $T(z)$ is an invariant of the outgoing scattering system $\mathfrak{S}^+$.

To get a complete set of unitary invariants we need an additional invariant defined as follows. We define an operator $L \colon L^2_{\text{fin}-}(\mathcal{F}_d \times \mathcal{F}_d, \mathcal{E}) \to L^2_{\text{arb}-}(\mathcal{F}_d \times \mathcal{F}_d, \mathcal{E})$ by

(5.1.6)

$$\left\langle L(S^{Rw} U^{R[*]v} f), S^{Rw'} U^{R[*]v'} f' \right\rangle_{L^2} = \left\langle \mathcal{S}^w \mathcal{S}^{*v} P_{\mathcal{R}} \Phi^{+*} f, \mathcal{S}^{w'} \mathcal{S}^{*v'} P_{\mathcal{R}} \Phi^{+*} f' \right\rangle_{\mathcal{R}}$$

for $f, f' \in L^2(\mathcal{F}_d \times \{\emptyset\}, \mathcal{E})$ and $v, w, v', w' \in \mathcal{F}_d$. For $w, v, w', v' = \emptyset$ and $f, f' \in L^2(\mathcal{F}_d, \mathcal{E})$ we have

$$\begin{aligned}\langle Lf, f'\rangle_{L^2} &= \langle P_{\mathcal{R}}\Phi^{+*}f, P_{\mathcal{R}}\Phi^{+*}f'\rangle_{\mathcal{R}} \\ &= \left\langle (I - P_{\mathcal{G}_*^+})\Phi^{+*}f, (I - P_{\mathcal{G}_*^+})\Phi^{+*}f'\right\rangle_{\mathcal{K}^+} \\ &= \langle \Phi^{+*}f, \Phi^{+*}f'\rangle_{\mathcal{K}} - \langle \mathbf{S}^+\Phi^{+*}f, \mathbf{S}^+\Phi^{+*}f'\rangle_{\mathcal{K}^+} \\ &= \langle f, f'\rangle_{L^2} - \langle M_T f, M_T f'\rangle_{L^2} \\ &= \langle (I - (M_T)^* M_T) f, f'\rangle_{L^2}\end{aligned}$$

and hence L is an extension of $I - (M_T)^* M_T$. By construction, the map

$$\Phi_{\mathcal{R}} \colon \mathcal{S}^w \mathcal{S}^{*v} P_{\mathcal{R}} f \mapsto L S^{Rw} U^{R[*]} f \tag{5.1.7}$$

is isometric from the linear manifold $\mathcal{R}_0$ to a dense subspace of $\mathcal{L}_L$ which also satisfies the intertwining conditions

$$\Phi_{\mathcal{R}}(\mathcal{S}_j|_{\mathcal{D}_0}) = \mathcal{U}_{L,j}\Phi_{\mathcal{R}}, \qquad \Phi_{\mathcal{R}}(\mathcal{S}_j^*|_{\mathcal{D}_0}) = \mathcal{U}_{L,j}^*\Phi_{\mathcal{R}} \tag{5.1.8}$$

where we define

$$\mathcal{U}_{L,j} \colon Lf \mapsto LS_j^R f, \qquad \mathcal{U}_{L,j}^* \colon Lf \mapsto LU_j^{R[*]} f.$$

Here we use that $U_j^{R[*]} S_k^R = \delta_{j,k} I$ which can easily be checked directly. Under the assumption that $\mathfrak{S}^+$ is minimal, $\mathcal{R}_0$ is dense in $\mathcal{R}$, and $\Phi_{\mathcal{R}}$ extends to a unitary map from $\mathcal{R}$ onto $\mathcal{L}_L$. Since $\mathcal{S}|_{\mathcal{R}}$ is row unitary, we conclude that $\mathcal{U}_L$ is row unitary. From the discussion in Section 2.2, we conclude that L is a Cuntz weight. In this way we have constructed a Cuntz weight L which extends the Cuntz-Toeplitz operator $I - (M_T)^* M_T$ and which is a unitary invariant for the outgoing scattering system $\mathfrak{S}^+$. Let us call L constructed in this way the *characteristic Cuntz weight* of $\mathfrak{S}^+$ and the pair (T, L) of invariants the *characteristic pair* for $\mathfrak{S}^+$.

In case $\mathfrak{S}^+$ is the forward part of a full scattering system $\mathfrak{S} = \mathfrak{S}(T, W(W_*)$, then we may write

$$\begin{aligned}\mathcal{S}^w \mathcal{S}^{*v} P_{\mathcal{R}} \Phi^{+*} f = \mathcal{U}^w \mathcal{U}^{*v} P_{\mathcal{R}} \Phi^{+*} f = P_{\mathcal{R}} \mathcal{U}^w \mathcal{U}^{*v} \Phi^* W f = P_{\mathcal{R}} \Phi^* \mathcal{U}_W^w \mathcal{U}_W^{*v} W f \\ = P_{\mathcal{R}} \Phi^* W S^{Rw} U^{R[*]v} f\end{aligned}$$

for $f \in L^2(\mathcal{F}_d, \mathcal{E})$, and hence

$$\begin{aligned}&\left\langle LS^{Rw} U^{R[*]v} f, S^{Rw'} U^{R[*]v'} f'\right\rangle_{L^2} := \left\langle \mathcal{U}^w \mathcal{U}^{*v} P_{\mathcal{R}} \Phi^{+*} f, \mathcal{U}^{w'} \mathcal{U}^{*v'} P_{\mathcal{R}} \Phi^{+*} f'\right\rangle_{\mathcal{R}} \\ &= \left\langle P_{\mathcal{R}} \Phi^* W S^{Rw} U^{R[*]v} f, P_{\mathcal{R}} \Phi^* W S^{Rw'} U^{R[*]v'} f'\right\rangle_{\mathcal{R}} \\ &= \left\langle (I - P_{\widetilde{\mathcal{G}}_*}) \Phi^* W S^{Rw} U^{R[*]v} f, (I - P_{\widetilde{\mathcal{G}}_*}) \Phi^* W S^{Rw'} U^{R[*]v'} f'\right\rangle_{\mathcal{K}} \\ &= \left\langle W S^{Rw} U^{R[*]v} f, W S^{Rw'} U^{R[*]v'} f'\right\rangle_{\mathcal{L}_W} \\ &\quad - \left\langle SW S^{Rw} U^{R[*]v} f, SW S^{Rw'} U^{R[*]v'} f'\right\rangle_{\mathcal{L}_{W_*}} \\ &= \left\langle W S^{Rw} U^{R[*]v} f, S^{Rw'} U^{R[*]v'} f'\right\rangle_{L^2} \\ &\quad - \left\langle W_* L_T S^{Rw} U^{R[*]v} f, L_T S^{Rw'} U^{R[*]v'} f'\right\rangle_{L^2}\end{aligned}$$

$$= \left\langle (W - L_T^{[*]} W_* L_T) S^{Rw} U^{R[*]v} f, S^{Rw'} U^{R[*]v'} f' \right\rangle_{L^2} \tag{5.1.9}$$

from which we conclude that $L = W - L_T^{[*]} W_* L_T$. The construction of L here explains why the formula (2.4.13) for the symbol $\widehat{L}(z, \zeta)$ of $L = W - L_T^{[*]} W_* L_T$ is independent of W_*.

From a characteristic pair (T, L), we can construct the Sz.-Nagy-Foiaş model for the outgoing Cuntz scattering system as follows. We are given (T, L), where $T(z) = \sum_{v \in \mathcal{F}_d} T_v z^v \in \mathcal{S}_{nc,d}(\mathcal{E}, \mathcal{E}_*)$ and where L is a Cuntz-weight extension of the Cuntz-Toeplitz operator $I - (M_T)^* M_T$. We define a model outgoing Cuntz scattering system

$$\mathfrak{S}^+_{NF'} = (\mathcal{S}_{NF'}; \mathcal{K}^+_{NF'}, \mathcal{G}^+_{NF'*}, \mathcal{G}_{NF'}) \tag{5.1.10}$$

given by

$$\mathcal{K}^+_{NF'} = \begin{bmatrix} L^2(\mathcal{F}_d, \mathcal{E}_*) \\ \mathcal{L}_L \end{bmatrix}, \qquad \mathcal{G}^+_{NF'*} = \begin{bmatrix} L^2(\mathcal{F}_d, \mathcal{E}_*) \\ 0 \end{bmatrix},$$

$$\mathcal{G}_{NF'} = \begin{bmatrix} M_T \\ L \end{bmatrix} L^2(\mathcal{F}_d, \mathcal{E}), \qquad \mathcal{S}_{NF'} = \begin{bmatrix} S^R & 0 \\ 0 & \mathcal{U}_L. \end{bmatrix}.$$

Then we have the following result.

THEOREM 5.1.1. *Suppose that (T, L) is a characteristic pair, i.e., $T(z) \in \mathcal{S}_{nc,d}(\mathcal{E}, \mathcal{E}_*)$ and L is a Cuntz-weight extension of the Cuntz-Toeplitz operator $I - (M_T)^* M_T$. Then $\mathfrak{S}^+ = \mathfrak{S}^+(T, L)$ defined as in (5.1.10) is a minimal outgoing scattering system with associated wandering subspaces and scattering subspace*

$$\mathcal{E}_{NF'*} = \begin{bmatrix} \mathcal{E}_* \\ 0 \end{bmatrix}, \qquad \mathcal{E}_{NF'} = \begin{bmatrix} T \\ L \end{bmatrix} \mathcal{E}, \qquad \mathcal{H}_{NF'} = \begin{bmatrix} L^2(\mathcal{F}_d, \mathcal{E}_*) \\ \mathcal{L}_L \end{bmatrix} \ominus \begin{bmatrix} M_T \\ L \end{bmatrix} L^2(\mathcal{F}_d, \mathcal{E})$$

and with characteristic pair $(T_{NF'}, L_{NF'})$ coinciding with (T, L) under the identification maps $i_{NF'} \colon \mathcal{E} \to \mathcal{E}_{NF'}$ and $i_{NF'*} \colon \mathcal{E}_* \to \mathcal{E}_{NF'*}$ given by

$$i_{NF'} \colon e \mapsto \begin{bmatrix} T \\ L \end{bmatrix} e, \qquad i_{NF'*} \colon e_* \mapsto \begin{bmatrix} e_* \\ 0 \end{bmatrix}.$$

Moreover, any other minimal outgoing Cuntz scattering system $\mathfrak{S}^+$ with characteristic pair (T, L) is unitarily equivalent to the Sz.-Nagy-Foiaş model outgoing scattering system $\mathfrak{S}^+_{NF'}(T, L)$ via the identification map $\mathcal{I}^+_{NF'} \colon \mathcal{K}^+ \to \mathcal{K}^+_{NF'}$ defined densely by

$$\mathcal{I}^+_{NF'} \colon \Phi_*^{++*} f + S^w S^{*v} P_{\mathcal{R}} \Phi^{++*} g \mapsto \begin{bmatrix} f \\ L(S^R)^w (U^{R[*]})^v g \end{bmatrix}$$

for $f \in L^2(\mathcal{F}_d, \mathcal{E}_*)$ and $g \in L^2(\mathcal{F}_d, \mathcal{E})$.

PROOF. It is a direct check that $\mathfrak{S}^+(T, L)$ is a minimal outgoing Cuntz scattering system with characteristic pair equal to (T, L) (after the identifications $i_{NF'}$ and $i_{NF'*}$ explained in the statement of the theorem). If $\mathfrak{S}^+$ is an arbitrary minimal outgoing Cuntz scattering system (5.1.1) with characteristic pair (T, L), then, as was observed above, minimality of $\mathfrak{S}^+$ is equivalent to the linear manifold $\mathcal{D}_0$ given by (5.1.4) being dense in $\mathcal{K}^+$. We thus see that $\mathcal{I}^+_{NF'}$ in the statement of the theorem is densely defined. We also know that $\mathcal{K}^+ = \mathcal{G}^+_* \oplus \mathcal{R}$ and that $\Phi^+_*|_{\mathcal{G}^+_*} \colon \mathcal{G}^+_* \to L^2(\mathcal{F}_d, \mathcal{E}_*)$ is unitary, and we have seen that the map $\Phi_{\mathcal{R}} \colon \mathcal{R} \to \mathcal{L}_L$

given by (5.1.7) is unitary (given that $\mathfrak{S}^+$ is minimal). Since $\mathcal{I}^+_{NF'} = \Phi^+_* \oplus \Phi_{\mathcal{R}}$ on the dense set $\mathcal{G}^+_* \oplus \mathcal{R}_0$ by definition, we see that $\mathcal{I}^+_{NF'} \colon \mathcal{K}^+ \to \mathcal{K}^+_{NF'}$ is unitary. It is routine to check also that

$$\mathcal{I}^+_{NF'}\mathcal{G} = \mathcal{G}_{NF'}, \qquad \mathcal{I}^+_{NF'}\mathcal{G}^+_* = \mathcal{G}^+_{NF'*}, \qquad \mathcal{I}^+_{NF'}\mathcal{S}_j = \mathcal{S}_{NF',j}\mathcal{I}^+_{NF'} \text{ for } j = 1, \dots, d.$$

This completes the proof of Theorem 5.1.1. □

5.2. Cuntz unitary colligations and outgoing Cuntz scattering systems

Let $\mathfrak{S}^+$ be an outgoing Cuntz scattering system as in (5.1.1). If we view $\mathfrak{S}^+$ as the forward part of a full Cuntz scattering system $\mathfrak{S}$ (3.1.1), we see that

$$\mathcal{H} = \mathcal{K} \ominus [\mathcal{G}_* \oplus \mathcal{G}] = \mathcal{K}^+ \ominus \mathcal{G}, \qquad \mathcal{E}_* = \mathcal{K}^+ \ominus \operatorname*{span}_{j=1,\dots,d} \mathcal{U}_j\mathcal{K}^+, \qquad \mathcal{E} = \mathcal{G} \ominus \operatorname*{span}_{j=1,\dots,d} \mathcal{U}_j\mathcal{G}$$

are all contained in $\mathcal{K}^+$, and the unitary colligation $U = U(\mathfrak{S}^+)$ as defined by (3.3.2) really depends only on the forward part $\mathfrak{S}^+$ of the scattering system $\mathfrak{S}$ ($U = U(\mathfrak{S}^+)$). This suggests that, given an outgoing Cuntz scattering system $\mathfrak{S}^+$ (5.1.1), we define an associated Cuntz unitary colligation $U = U(\mathfrak{S}^+)$ by

$$U = \begin{bmatrix} A & B \\ C & D \end{bmatrix} = \begin{bmatrix} A_1 & B_1 \\ \vdots & \vdots \\ A_d & B_d \\ C & D \end{bmatrix} \colon \begin{bmatrix} \mathcal{H} \\ \mathcal{E} \end{bmatrix} \to \begin{bmatrix} \oplus_{j=1}^d \mathcal{H} \\ \mathcal{E}_* \end{bmatrix} \tag{5.2.1}$$

with

$$\begin{aligned} \begin{bmatrix} A_j & B_j \end{bmatrix} &= P_{\mathcal{H}}\mathcal{S}_j^*|_{\mathcal{H}\oplus\mathcal{E}}, \\ \begin{bmatrix} C & D \end{bmatrix} &= P_{\mathcal{E}_*}|_{\mathcal{H}\oplus\mathcal{E}}. \end{aligned} \tag{5.2.2}$$

where $\mathcal{H} := \mathcal{K}^+ \ominus \mathcal{G}$ is the scattering space for the outgoing Cuntz scattering system $\mathcal{K}^+$. Just as in Section 3.3, even though $\mathcal{S}$ is merely a row isometry rather than a row unitary, one can show that U so defined is unitary, and that $\mathfrak{S}^+$ is minimal if and only if U is closely connected.

Furthermore, if we define forward-side time-domain Fourier representation operators

$$\begin{aligned} \widetilde{\Phi}^+ &\colon k \mapsto \{u(v)\}_{v\in\mathcal{F}_d} \text{ where } u(v) = P_{\mathcal{E}}\mathcal{S}^{*v}k, \\ \widetilde{\Phi}^+_* &\colon k \mapsto \{y(v)\}_{v\in\mathcal{F}_d} \text{ where } y(v) = P_{\mathcal{E}_*}\mathcal{S}^{*v}k, \\ \widetilde{\Phi}^+_{\mathcal{H}} &\colon k \mapsto \{x(v)\}_{v\in\mathcal{F}_d} \text{ where } x(v) = P_{\mathcal{H}}\mathcal{S}^{*v}k, \end{aligned}$$

and if we let Ω^+ denote the operator encoding the whole aggregate of these

$$\Omega^+ \colon k \mapsto (u, x, y) := (\widetilde{\Phi}^+ k, \widetilde{\Phi}^+_{\mathcal{H}} k, \widetilde{\Phi}^+_* k), \tag{5.2.3}$$

then we have the following analogue of Theorem 3.3.7.

THEOREM 5.2.1. *For $\mathfrak{S}^+$ an outgoing Cuntz scattering system as in* (5.1.1) *and Ω^+ defined as in* (5.2.3), *then, for any $k \in \mathcal{K}$, $(u, x, y) = \Omega^+ k$ satisfies the forward system equations associated with the Cuntz unitary colligation $U = U(\mathfrak{S}^+)$* (5.2.2):

$$\begin{aligned} x(g_j v) &= A_j x(v) + B_j u(v) \\ y(v) &= Cx(v) + Du(v) \end{aligned} \tag{5.2.4}$$

Conversely, any trajectory (u, x, y) of (5.2.4) *with $u \in \ell^2(\mathcal{F}_d, \mathcal{E})$ is of the form $(u, x, y) = \Omega^+ k$ for a $k \in \mathcal{K}^+$.*

PROOF. For purposes of the proof, we may consider $\mathfrak{S}^+$ as being the forward part of a full Cuntz scattering system $\mathfrak{S}$. Then we simply quote Theorem 3.3.7 and restrict all trajectories there to $\mathcal{F}_d \times \{\emptyset\}$. □

In analogy with Section 4.2, we may start with a Cuntz unitary colligation (5.2.1) and define the space $\mathcal{T}_U^+$ to consist of all trajectories (u, x, y) (defined on $\mathcal{F}_d$ with values in $\mathcal{E}$, $\mathcal{H}$ and $\mathcal{E}_*$ respectively) of the forward system equations (5.2.4) such that $u \in \ell^2(\mathcal{F}_d, \mathcal{E})$ with norm

$$\|(u, x, y)\|_{\mathcal{T}^+}^2 = \|u\|_{\ell^2(\mathcal{F}_d, \mathcal{E})} + \|x(\emptyset)\|_{\mathcal{H}}^2. \tag{5.2.5}$$

By restricting the formulas (4.1.9) and (4.1.10) to the domain $\mathcal{F}_d \times \{\emptyset\}$, we see that y and $x(v)$ for $v \neq \emptyset$ are uniquely determined by u and $x(\emptyset)$ according to the formulas

$$x^{\wedge +}(z) = (I - Z_r(z)A)^{-1}[x(\emptyset) + Z_r(z)Bu^{\wedge +}(z)]$$
$$y^{\wedge +}(z) = C(I - Z_r(z)A)^{-1}x(\emptyset) + T(z)u^{\wedge +}(z)$$

(where $T(z) = T_{\Sigma(U)}(z) = D + C(I - Z_r(z)A)^{-1}Z_r(z)B$ is the characteristic function of the colligation U), and that $x(\emptyset) \in \mathcal{H}$ and $u \in \ell^2(\mathcal{F}_d, \mathcal{E})$ can be taken to be arbitrary. It follows that $\mathcal{T}^+$ is a Hilbert space in the $\mathcal{T}_U^+$-norm, and that the map

$$\mathcal{I}_U^+ \colon (u, x, y) \mapsto \begin{bmatrix} x(\emptyset) \\ u^{\wedge +}(z) \end{bmatrix} \tag{5.2.6}$$

is a unitary map from $\mathcal{T}_U^+$ onto $\mathcal{H} \oplus L^2(\mathcal{F}_d, \mathcal{E})$. Furthermore, by restricting the results in Theorem 4.2.6 to the forward part $\mathcal{F}_d \times \{\emptyset\}$, we arrive at the next theorem. To state the theorem, we define a d-tuple of operators $\mathcal{S}_U = (\mathcal{S}_{U,1}, \ldots, \mathcal{S}_{U,d})$ on $\mathcal{T}_U^+$ by

$$\mathcal{S}_{U,j} \colon (u, x, y) \mapsto (S_j^R u, S_j^R x, S_j^R y). \tag{5.2.7}$$

It turns out that $\mathcal{S}_U$ is a row isometry, and hence has a Wold decomposition

$$\mathcal{T}_U^+ = \mathcal{G}_{U;*}^+ \oplus \mathcal{R}_U$$

where

$$\mathcal{G}_{U*}^+ = \oplus_{v \in \mathcal{F}_d} \mathcal{S}_U^v \mathcal{E}_{U;*} \tag{5.2.8}$$

where

$$\mathcal{E}_{U;*} = \mathcal{T}_U^+ \ominus \operatorname*{span}_{j=1,\ldots,d} \mathcal{S}_{U,j}\mathcal{T}_U^+. \tag{5.2.9}$$

In addition define the subspace

$$\mathcal{G}_U = \{(u, x, y) \in \mathcal{T}_U^+ : x(\emptyset) = 0\} \tag{5.2.10}$$

and let $\mathfrak{S}_U^+$ be the collection

$$\mathfrak{S}^+(U) = (\mathcal{S}_U = (\mathcal{S}_{U,1}, \ldots, \mathcal{S}_{U,d}); \mathcal{T}_U^+, \mathcal{G}_U, \mathcal{G}_{U*}^+). \tag{5.2.11}$$

THEOREM 5.2.2. *Let U be a Cuntz unitary colligation* (5.2.1) *with admissible space of trajectories*

$$\mathcal{T}_U^+ = \{(u, x, y) \colon (u, x, y) \text{ solves } (5.2.4)\ , u \in \ell^2(\mathcal{F}_d, \mathcal{E})\}$$

with $\mathcal{T}_U^+$-norm (5.2.5). *Then $\mathcal{S}_U$ defined by* (5.2.7) *is a row isometry on $\mathcal{T}_U^+$ and the system $\mathfrak{S}_U^+$ defined by* (5.2.11) *is an outgoing Cuntz scattering system with outgoing wandering subspace $\mathcal{E}_U$ and incoming wandering subspace $\mathcal{E}_{U;*}$ and scattering subspace $\mathcal{H}_U$ given explicitly by*

$$\mathcal{E}_U = \{(u,x,y) \in \mathcal{T}_U^+ : x(\emptyset) = 0 \text{ and } u|_{\mathcal{F}_d \setminus \{\emptyset\}} = 0\}, \tag{5.2.12}$$

$$\mathcal{E}_{U;*} = \{(u,x,y) \in \mathcal{T}_U^+ : A_j x(\emptyset) + B_j u(\emptyset) = 0 \text{ for } j = 1, \dots, d \text{ and } u|_{\mathcal{F}_d \setminus \{\emptyset\}} = 0\}, \tag{5.2.13}$$

$$\mathcal{H}_U = \{(u,x,y) \in \mathcal{T}_U^+ : u = 0\}. \tag{5.2.14}$$

Moreover, under the unitary identification maps $\iota \colon \mathcal{E}_U \to \mathcal{E}$, $\iota_* \colon \mathcal{E}_{U;*} \to \mathcal{E}_*$ and $\iota_{\mathcal{H}} \colon \mathcal{H}_U \to \mathcal{H}$ given by

$$\iota \colon (u,x,y) \in \mathcal{E}_U \mapsto u(\emptyset) \in \mathcal{E},$$

$$\iota_* \colon (u,x,y) \in \mathcal{E}_{U;*} \mapsto Cx(\emptyset) + Du(\emptyset) \in \mathcal{E}_*,$$

$$\iota_{\mathcal{H}} \colon (u,x,y) \in \mathcal{H}_U \mapsto x(\emptyset) \in \mathcal{H},$$

the original Cuntz unitary colligation U coincides with the Cuntz unitary colligation $U(\mathfrak{S}_U^+)$ constructed from $\mathfrak{S}_U^+$ as in (5.2.2)*:*

$$U = \begin{bmatrix} \iota_{\mathcal{H}} & 0 \\ 0 & \iota_* \end{bmatrix} U(\mathfrak{S}_U^+) \begin{bmatrix} \iota_{\mathcal{H}}^* & 0 \\ 0 & \iota^* \end{bmatrix}.$$

The proof of Theorem 5.2.2 actually goes through a more explicit coordinate representation for $\mathcal{S}_U$ on $\mathcal{H} \oplus \ell^2(\mathcal{F}_d, \mathcal{E})$ which is more convenient for explicit computations. For this purpose, define a d-tuple of operators $\boldsymbol{\mathcal{S}}_U = (\boldsymbol{\mathcal{S}}_{U;1}, \dots, \boldsymbol{\mathcal{S}}_{U;d})$ on $\boldsymbol{\mathcal{K}}_U^+ := \mathcal{H} \oplus L^2(\mathcal{F}_d, \mathcal{E})$ by

$$\boldsymbol{\mathcal{S}}_{U;j} \colon \begin{bmatrix} h \\ g \end{bmatrix} \mapsto \begin{bmatrix} h' \\ g' \end{bmatrix} \tag{5.2.15}$$

where

$$h' = A_j^* h, \qquad g' = B_j^* h + g(z) z_j.$$

Then restriction of Theorem 4.2.3 to the forward time axis $\mathcal{F}_d \times \{\emptyset\}$ gives the following theorem.

THEOREM 5.2.3. *The d-tuple of operators $\boldsymbol{\mathcal{S}}_U = (\boldsymbol{\mathcal{S}}_{U;1}, \dots, \boldsymbol{\mathcal{S}}_{U;d})$ defined by* (5.2.15) *is a row isometry on the Hilbert space $\boldsymbol{\mathcal{K}}_U^+ := \mathcal{H} \oplus L^2(\mathcal{F}_d, \mathcal{E})$. If we set*

$$\boldsymbol{\mathcal{G}}_{U;*}^+ = \oplus_{v \in \mathcal{F}_d} \boldsymbol{\mathcal{S}}^v \boldsymbol{\mathcal{E}}_*$$

where

$$\begin{aligned} \boldsymbol{\mathcal{E}}_* :=& \boldsymbol{\mathcal{K}}^+ \ominus \operatorname*{span}_{j=1,\dots,d} \boldsymbol{\mathcal{S}}_j \boldsymbol{\mathcal{K}}^+ \\ =& \left\{ \begin{bmatrix} h \\ f(z) \end{bmatrix} \in \boldsymbol{\mathcal{H}}_U \colon f(z) = f_\emptyset z^\emptyset \text{ and } A_j h + B_j f_\emptyset = 0 \text{ for } j = 1, \dots, d \right\} \end{aligned}$$

and let $\boldsymbol{\mathfrak{S}}_U^+$ be the collection

(5.2.16)
$$\boldsymbol{\mathfrak{S}}^+ = \left(\boldsymbol{\mathcal{S}} = (\boldsymbol{\mathcal{S}}_1, \dots, \boldsymbol{\mathcal{S}}_d); \boldsymbol{\mathcal{K}}^+ = \begin{bmatrix} \mathcal{H} \\ L^2(\mathcal{F}_d, \mathcal{E}) \end{bmatrix}, \boldsymbol{\mathcal{G}}_U = \begin{bmatrix} 0 \\ L^2(\mathcal{F}_d, \mathcal{E}) \end{bmatrix}, \boldsymbol{\mathcal{G}}_{U;*}^+ \right),$$

then $\boldsymbol{\mathfrak{S}}^+$ is an outgoing Cuntz scattering system unitarily equivalent to the outgoing Cuntz scattering system $\mathfrak{S}^+(U)$ given in (5.2.11) *via the unitary map $\mathcal{I}_U \colon \mathcal{T}_U^+ \to$*

$\boldsymbol{\mathcal{K}}_U^+$ given by (5.2.6). Moreover, under the unitary identification maps $\iota\colon \boldsymbol{\mathcal{E}}_U \to \mathcal{E}$, $\iota_*\colon \boldsymbol{\mathcal{E}}_{U;*} \to \mathcal{E}_*$ and $\iota_{\mathcal{H}}\colon \boldsymbol{\mathcal{H}}_U \to \mathcal{H}$ for this case given by

$$\iota\colon \begin{bmatrix} 0 \\ ez^{\emptyset} \end{bmatrix} \in \mathcal{E}_U \mapsto e \in \mathcal{E},$$

$$\iota_*\colon \begin{bmatrix} h \\ ez^{\emptyset} \end{bmatrix} \in \mathcal{E}_{U;*} \mapsto Ch + De \in \mathcal{E}_*,$$

$$\iota_{\mathcal{H}}\colon \begin{bmatrix} h \\ 0 \end{bmatrix} \in \mathcal{H}_U \mapsto h \in \mathcal{H},$$

the original Cuntz unitary colligation U coincides with the Cuntz unitary colligation $U(\boldsymbol{\mathfrak{S}}_U^+)$ constructed from $\boldsymbol{\mathfrak{S}}_U^+$ as in (5.2.2):

$$U = \begin{bmatrix} \iota_{\mathcal{H}} & 0 \\ 0 & \iota_* \end{bmatrix} U(\boldsymbol{\mathfrak{S}}_U^+) \begin{bmatrix} \iota_{\mathcal{H}}^* & 0 \\ 0 & \iota^* \end{bmatrix}.$$

From Proposition 4.2.11 we see how to compute the characteristic pair (T, L) for $\mathfrak{S}_U^+$ (or $\boldsymbol{\mathfrak{S}}_U^+$) in terms of $U = \begin{bmatrix} A & B \\ C & D \end{bmatrix}$.

THEOREM 5.2.4. *Let $U = \begin{bmatrix} A & B \\ C & D \end{bmatrix}$ be a Cuntz unitary colligation (5.2.1) with associated outgoing Cuntz scattering system $\mathfrak{S}_U^+$ or $\boldsymbol{\mathfrak{S}}_U^+$ ((5.2.11) and (5.2.16)). Then the characteristic pair (T, L) for $\mathfrak{S}_U^+$ or $\boldsymbol{\mathfrak{S}}_U^+$ coincides with*

$$T(z) = D + C(I - Z_r(z)A)^{-1}Z_r(z)B, \tag{5.2.17}$$

$$L_{v,w;\alpha,\beta} = \begin{cases} \widehat{L}_{(v\alpha^{-1})\beta^\top, w} & \text{if } |v| \geq |\alpha|, \\ \widehat{L}_{\beta^\top, w(\alpha v^{-1})^\top} & \text{if } |v| \leq |\alpha| \end{cases} \tag{5.2.18}$$

where $\widehat{L}(z, \zeta) = \sum_{v,w} \widehat{L}_{v,w} z^v \zeta^w$ is given by

$$\begin{aligned} \widehat{L}(z, \zeta) = {} & B^* Z_r(\zeta)^* (I - A^* Z_r(\zeta)^*)^{-1} (I - Z_r(z)A)^{-1} Z_r(z) B \\ & + I - T(\zeta)T(z) - T(\zeta)^* k_{per}(z, \zeta) T(z) \\ & \text{with } T(z) \text{ as in (5.2.17).} \end{aligned} \tag{5.2.19}$$

PROOF. For purposes of the proof, let T' denote the characteristic function of the outgoing Cuntz scattering system $\mathfrak{S}_U^+$. Pick any Cuntz-weight extension of the identity W_* and consider the outgoing Cuntz scattering system as the forward part of a full Cuntz scattering system ($\boldsymbol{\mathfrak{S}}_{U,W_*}$ as in (4.2.14) or $\mathfrak{S}_{U,W_*}$ as in (4.2.24)). Then, as explained in Remark 4.2.4 (see also Remark 4.2.7), the scattering function $S = \Phi_* \Phi^*$ for $\mathfrak{S}_{U,W_*}$ coincides with $S = L_T^{W,W_*}$. On the other hand the characteristic function T' for $\mathfrak{S}^+(U)$ is equal to $\Phi_*^+ \Phi^{+*}$. We now note the connection between Φ and Φ^+:

$$\Phi|_{\mathcal{G}} = W\Phi^+|_{\mathcal{G}}, \qquad \Phi^* W|_{L^2(\mathcal{F}_d \times \{\emptyset\}, \mathcal{E})} = \Phi^{+*}$$

and similarly for Φ_* and Φ_*^+. (This follows from the general fact that $[Wf]_{v,\emptyset} = f_v$ for $f(z) = \sum_{v \in \mathcal{F}_d} f_v z^v \in L^2(\mathcal{F}_d, \mathcal{E})$ which in turn follows from the fact that W is a shift Cuntz weight.) Hence, we compute, for $f \in L^2(\mathcal{F}_d, \mathcal{E})$,

$$\begin{aligned} W_* T f &= SWf = \Phi_* \Phi^* W f = \Phi_* \Phi^{+*} f = \Phi_* P_{\mathcal{G}_*^+} \Phi^{+*} f = W_* \Phi_*^+ P_{\mathcal{G}_*^+} \Phi^{+*} f \\ &= W_* \Phi_*^+ \Phi^{+*} f = W_* T' f \end{aligned}$$

and we conclude that $T = T'$ as wanted.

We next verify formula (5.2.19). We know that the full Cuntz scattering system $\mathfrak{S}_{U,W_*}$ is alternatively determined (up to unitary equivalence, assuming minimality) by the triple of parameters (T, W, W_*) in the scattering function $S = L_T^{W,W_*}$ rather than by the parameters U, W_*, where T is the characteristic function (5.2.17) of U and where the symbol $\widehat{W}(z, \zeta)$ of the outgoing Cuntz weight W is given in terms of W_* and T by (4.2.5). By the computation (5.1.9) we know that the characteristic Cuntz weight for the outgoing Cuntz scattering system $\mathfrak{S}^+(U)$ is given as $L = W - L_T^{[*]} W_* L_T$ (with the dependence on the choice of W_* only apparent, since the definition (5.1.6) of L does not involve W_*). The symbol $\widehat{L}(z, \zeta)$ for $L = W - L_T^{[*]} W_* L_T$ is given by (4.2.52) in Proposition 4.2.11. This leads us to the formula (5.2.19), where the apparent dependence on W_* has disappeared as was to be expected. $\square$

5.3. Model theory for row contractions

Let us suppose that we start with a *characteristic pair* (T, L), i.e., a formal power series $T(z) \in \mathcal{S}_{nc,d}(\mathcal{E}, \mathcal{E}_*)$ together with a Cuntz-weight extension L of the Cuntz-Toeplitz operator $I - (M_T)^* M_T$. In Section 5.1 we associate with (T, L) the Sz.-Nagy-Foiaş model for an outgoing Cuntz scattering system $\mathfrak{S}^+_{NF'}$ (5.1.10). In this section we adopt the point of view of operator model theory as in [**NaF70, Po89a, Po89b**] and focus on the scattering subspace $\mathcal{H}_{NF'} := \mathcal{K}^+_{NF'} \ominus \mathcal{G}_{NF'}$ to get a model for row contractions.

Given a characteristic pair (T, L), we define the *Sz.-Nagy-Foiaş functional model* (for a row contraction) to be the space

$$\mathcal{H}_{(T,L)} := \begin{bmatrix} L^2(\mathcal{F}_d, \mathcal{E}_*) \\ \mathcal{L}_L \end{bmatrix} \ominus \begin{bmatrix} M_T \\ L \end{bmatrix} L^2(\mathcal{F}_d, \mathcal{E})$$

together with the d-tuple of operators $\mathbf{T}_{T,L} = (\mathbf{T}_{T,L;1}, \dots, \mathbf{T}_{T,L;d})$ given by

$$\mathbf{T}_{T,L;j} = P_{\mathcal{H}_{T,L}} \begin{bmatrix} S_j^R & 0 \\ 0 & U_{L,j} \end{bmatrix} \Big|_{\mathcal{H}_{T,L}} \quad \text{for } j = 1, \dots, d.$$

As the d-tuple

$$\left(\begin{bmatrix} S_1^R & 0 \\ 0 & U_{L,1} \end{bmatrix}, \dots, \begin{bmatrix} S_d^R & 0 \\ 0 & U_{L,d} \end{bmatrix} \right)$$

is a row isometry on $\begin{bmatrix} L^2(\mathcal{F}_d, \mathcal{E}_*) \\ \mathcal{L}_L \end{bmatrix}$, we see that $\mathbf{T}_{(T,L)}$ is a row contraction, i.e., the operator block-row matrix $\begin{bmatrix} \mathbf{T}_{T,L;1} & \cdots & \mathbf{T}_{T,L;d} \end{bmatrix} : \oplus_{j=1}^d \mathcal{H}_{T,L} \to \mathcal{H}_{T,L}$ is a contraction.

Conversely, suppose that $\mathbf{T} = (\mathbf{T}_1, \dots, \mathbf{T}_d)$ is a row contraction on a Hilbert space $\mathcal{H}$. Then we set A equal to the contraction operator given by

$$A = \begin{bmatrix} A_1 \\ \vdots \\ A_d \end{bmatrix} : \mathcal{H} \to \begin{bmatrix} \mathcal{H} \\ \vdots \\ \mathcal{H} \end{bmatrix} \quad \text{where } A_j = \mathbf{T}_j^* \text{ for } j = 1, \dots, d.$$

We next use the Halmos unitary-dilation procedure to embed A into a Cuntz unitary colligation

$$U = \begin{bmatrix} A & B \\ C & D \end{bmatrix} = \begin{bmatrix} A_1 & B_1 \\ \vdots & \vdots \\ A_d & B_d \\ C & D \end{bmatrix} : \begin{bmatrix} \mathcal{H} \\ \mathcal{E} \end{bmatrix} \to \begin{bmatrix} \mathcal{H} \\ \vdots \\ \mathcal{H} \\ \mathcal{E}_* \end{bmatrix}$$

where

$$B = D_T := \left(\begin{bmatrix} I & & \\ & \ddots & \\ & & I \end{bmatrix} - \begin{bmatrix} \mathbf{T}_1^* \\ \vdots \\ \mathbf{T}_d^* \end{bmatrix} \begin{bmatrix} \mathbf{T}_1 & \dots & \mathbf{T}_d \end{bmatrix} \right)^{1/2} : \mathcal{D}_{\mathbf{T}} \to \oplus_{j=1}^d \mathcal{H},$$

$$C = D_{\mathbf{T}^*} = (I - \mathbf{T}_1\mathbf{T}_1^* - \cdots - \mathbf{T}_d\mathbf{T}_d^*)^{1/2} \colon \mathcal{H} \to \mathcal{D}_{\mathbf{T}^*},$$

$$D = -\begin{bmatrix} \mathbf{T}_1 & \dots & \mathbf{T}_d \end{bmatrix} |_{\mathcal{D}_{\mathbf{T}}} \colon \mathcal{D}_{\mathbf{T}} \to \mathcal{D}_{\mathbf{T}^*}$$

where we have set

$$\mathcal{E} = \mathcal{D}_{\mathbf{T}} := \text{clos. im}\, D_{\mathbf{T}} \subset \oplus_{j=1}^d \mathcal{H}, \qquad \mathcal{E}_* = \mathcal{D}_{\mathbf{T}^*} := \text{clos. im}\, D_{\mathbf{T}^*} \subset \mathcal{H}.$$

Unitary colligations $U = \begin{bmatrix} A & B \\ C & D \end{bmatrix}$ arising in this way, i.e., as the Halmos dilation of a contraction operator $A = \mathbf{T}^*$, can be characterized as those which are *strict* in the sense of Livšic (see [**Bro71, Li73**]), i.e., as those for which both B and C^* are injective. The corresponding restriction on the characteristic function $T(z) = D + C(I - Z_r(z)A)^{-1}Z_r(z)B$ is that $T(z)$ be *pure* in the sense of Sz.-Nagy and Foiaş [**NaF70**], i.e., that there be no nonzero subspace of $\mathcal{E}$ on which D is isometric.

We may then embed U into an outgoing Cuntz scattering system $\mathfrak{S}_U^+$ (5.2.11), or, in more explicit form in terms of coordinates, $\boldsymbol{\mathfrak{S}}_U^+$ (5.2.16). The operator d-tuple $\boldsymbol{\mathcal{S}} = (\boldsymbol{\mathcal{S}}_{U;1}, \dots, \boldsymbol{\mathcal{S}}_{U;d})$ appearing in $\boldsymbol{\mathfrak{S}}_U^+$ (see (5.2.15)) is exactly equal to the minimal row-isometric dilation for $\mathbf{T}$ as constructed by Popescu [**Po89c**]. We then define the *characteristic pair* $(T, L) = (T_{\mathbf{T}}, L_{\mathbf{T}})$ for the row contraction $\mathbf{T}$ to consist of the characteristic function $T(z)$ and the characteristic Cuntz weight L for the outgoing Cuntz scattering system $\mathfrak{S}_U^+$ or $\boldsymbol{\mathfrak{S}}_U^+$ (with the appropriate identifications of the outgoing and incoming wandering subspaces with $\mathcal{E}$ and $\mathcal{E}_*$ respectively). Explicitly, we therefore have

$$T_{\mathbf{T}}(z) = (-\mathbf{T} + D_{\mathbf{T}^*}(I - Z_r(z)\mathbf{T}^*)^{-1}Z_r(z)D_{\mathbf{T}})|_{\mathcal{D}_{\mathbf{T}}} \colon \mathcal{D}_{\mathbf{T}} \to \mathcal{D}_{\mathbf{T}^*} \tag{5.3.1}$$

$$[L_{\mathbf{T}}]_{v,w;\alpha,\beta} = \begin{cases} \widehat{L}_{\mathbf{T};(v\alpha)^{-1})\beta^\top,w} & \text{if } |v| \geq |\alpha|, \\ \widehat{L}_{\mathbf{T};\beta^\top,w(\alpha v^{-1})^\top} & \text{if } |v| \leq |\alpha| \end{cases} \tag{5.3.2}$$

where $\widehat{L}_{\mathbf{T}}(z,\zeta) = \sum_{v,w} \widehat{L}_{\mathbf{T};v,w} z^v \zeta^w$ is given by

$$\begin{aligned} \widehat{L}_{\mathbf{T}}(z,\zeta) = &[D_{\mathbf{T}} Z_r(\zeta)^*(I - \mathbf{T}Z_r(\zeta)^*)^{-1}(I - Z_r(z)\mathbf{T}^*)^{-1}D_{\mathbf{T}} \\ &+ I - T(\zeta)^*T(z) - T(\zeta)^*k_{per}(z,\zeta)T(z)] \colon \mathcal{D}_{\mathbf{T}} \to \mathcal{D}_{\mathbf{T}} \\ &\text{with } T(z) = T_{\mathbf{T}}(z) \text{ as in (5.3.1) and } k_{per} \text{ as in (2.2.26).} \end{aligned}$$

(Here we use formula (2.4.26) combined with (4.2.51) or directly (4.2.52) with A, B, C, D equal to the Halmos dilation of $\mathbf{T}^*$). When we combine this discussion with the results of Sections 5.1 and 5.2 (especially Theorem 5.1.1), we arrive at the following result.

THEOREM 5.3.1. *Let $\mathbf{T} = (\mathbf{T}_1, \dots, \mathbf{T}_d)$ be a completely nonunitary row contraction on the (separable) Hilbert space $\mathcal{H}$ ($\neq \{0\}$) with characteristic pair $(T, L) = (T_{\mathbf{T}}, L_{\mathbf{T}})$. Then $\mathbf{T}$ is unitarily equivalent to the operator d-tuple*

$$\mathbf{T}_{T,L} = (\mathbf{T}_{T,L;1} \dots, \mathbf{T}_{T,L;d})$$

on the functional Hilbert space

$$\mathcal{H}_{T,L} := \mathcal{H}_{NF'} = \begin{bmatrix} L^2(\mathcal{F}_d, \mathcal{D}_{\mathbf{T}^*}) \\ \mathcal{L}_L \end{bmatrix} \ominus \begin{bmatrix} T \\ L \end{bmatrix} L^2(\mathcal{F}_d, \mathcal{D}_{\mathbf{T}})$$

defined by

$$\mathbf{T}_{T,L;j} = P_{\mathcal{H}(T,L)} \begin{bmatrix} S_j^R & 0 \\ 0 & \mathcal{U}_{L,j} \end{bmatrix} \Bigg|_{\mathcal{H}(T,L)}.$$

Conversely, given a characteristic pair (T, L), the associated d-tuple of operators $\mathbf{T}_{(T,L)} = (\mathbf{T}_{T,L;1} \dots, \mathbf{T}_{T,L;d})$ has characteristic pair (T', L') which coincides with (T, L). In general if (T, L) and (T', L') are two characteristic pairs which coincide, then the associated model operator d-tuples $\mathbf{T}_{T,L}$ and $\mathbf{T}_{T',L'}$ are unitarily equivalent.

REMARK 5.3.2. If we consider $\mathbf{T}$ on $\mathcal{H}$ as embedded in its row-isometric dilation $\mathcal{S}$ acting on $\mathcal{K}^+ \supset \mathcal{H}$, then the mapping implementing the unitary equivalence between $\mathbf{T}$ and $\mathbf{T}_{T,L}$ in Theorem 5.3.1 is simply the restriction of $\mathcal{I}'_{NF'}$ to $\mathcal{H} \subset \mathcal{K}^+$, where $\mathcal{I}'_{NF'}$ is the identification operator appearing in Theorem 5.1.1. Note that the first component of $\mathcal{I}'_{NF'}$ amounts to Φ_*^+ while the formula for the restriction of Φ_* to $\mathcal{H}$ amounts to the formula for $\widehat{y}(z, \zeta)$ in (4.2.4) with $y_\perp = 0$ and $u_+ = 0$:

$$\Phi_* \colon h \mapsto W_*(z, \zeta) C(I - Z_r(z)A)^{-1} h.$$

The second component of $\mathcal{I}'_{NF'}$ can be computed as $P_{\mathcal{R}} h = h - \sum_{v \in \mathcal{F}_d} \mathcal{S}^v (P_{\mathcal{E}} \mathcal{S}^{*v} h)$. Therefore, the action of $\mathcal{I}'_{NF'}$ on $\mathcal{H}$ can be given somewhat more explicitly as

$$\mathcal{I}'_{NF'} \colon h \mapsto \begin{bmatrix} C(I - Z(z)A)^{-1} h \\ \Phi_{\mathcal{R}} \left(h - \sum_{v \in \mathcal{F}_d} \mathcal{S}^v (P_{\mathcal{E}_*} \mathcal{S}^{*v} h) \right) \end{bmatrix} \tag{5.3.3}$$

where $\Phi_{\mathcal{R}}$ is the unitary transformation from $\mathcal{R}$ to $\mathcal{L}_L$ defined densely by (5.1.7).

The L_2-norm of the first component in (5.3.3) can be computed explicitly as

$$\begin{aligned} \|C(I - Z_r(z)A)^{-1} h\|_{L^2}^2 &= \sum_{n=0}^{\infty} \left[\sum_{v \colon |v| = n} \|CA^v h\|^2 \right] \\ &= \sum_{n=0}^{\infty} \left[\sum_{v \colon |v| = n} \langle A^{*v^\top} C^* C A^v h, h \rangle \right] \\ &= \sum_{n=0}^{\infty} \left[\sum_{v \colon |v| = n} A^{*v^\top} (I - A_1^* - \dots - A_d^* A) A^v h, h \rangle \right] \\ &= \|h\|^2 - \lim_{N \to \infty} \left[\sum_{v \colon |v| = N} \|A^v h\|^2 \right]. \end{aligned}$$

Hence we see that $\Phi_*|_{\mathcal{H}} \colon \mathcal{H} \to L^2(\mathcal{F}_d, \mathcal{E})$ is isometric if and only if $A = \mathbf{T}^*$ is *d-stable* in the sense of (4.1.26) (corresponding to the $C_{\cdot 0}$ case in the terminology of [**NaF70**]). By Proposition 4.1.3, this is exactly the same condition as that required for M_T to be isometric (or, in the terminology of [**NaF70**], for T to be *inner*).

Hence in this case, we have $I - (M_T)^* M_T = 0$ from which we get $L = 0$ and $\mathcal{R} = 0$. This leads to the simplification of the model space

$$\mathcal{H}_{(T,L)} = L^2(\mathcal{F}_d, \mathcal{D}_{\mathbf{T}^*}) \ominus T L^2(\mathcal{F}_d, \mathcal{D}_{\mathbf{T}})$$

and the row-isometric dilation space $\mathcal{K}^+_{NF'}$ simplifies to

$$\mathcal{K}^+_{NF'} = L^2(\mathcal{F}_d, \mathcal{D}_{T^*}) = \mathcal{H}_{T,L} \oplus T L^2(\mathcal{F}_d, \mathcal{D}_{\mathbf{T}}).$$

As a corollary we have the following.

Corollary 5.3.3. *Two completely nonunitary row contractions $\mathbf{T} = (\mathbf{T}_1, \dots, \mathbf{T}_d)$ and $\mathbf{T}' = (\mathbf{T}'_1, \dots, \mathbf{T}'_d)$ are unitarily equivalent if and only if their characteristic pairs $(T_{\mathbf{T}}, L_{\mathbf{T}})$ and $(T_{\mathbf{T}'}, L_{\mathbf{T}'})$ coincide.*

Proof. If $\mathbf{T}$ and $\mathbf{T}'$ are unitarily equivalent, we can use the formulas (5.3.1) and (5.3.2) to see directly that the associated characteristic pairs $(T_{\mathbf{T}}, L_{\mathbf{T}})$ and $(T_{\mathbf{T}'}, L_{\mathbf{T}'})$ coincide.

Conversely, suppose that $(T, L) := (T_{\mathbf{T}}, L_{\mathbf{T}})$ and $(T', L') := (T_{\mathbf{T}'}, L_{\mathbf{T}'})$ coincide. Then it is a direct verification (also a part of the statement of Theorem 5.3.1) that the associated model row contractions $\mathbf{T}_{T,L}$ and $\mathbf{T}_{T',L'}$ are unitarily equivalent. But, by Theorem 5.3.1, $\mathbf{T}$ is unitarily equivalent to $\mathbf{T}_{T,L}$ and $\mathbf{T}'$ is unitarily equivalent to $\mathbf{T}_{T',L'}$. We conclude that $\mathbf{T}$ is unitarily equivalent to $\mathbf{T}'$ as claimed.

One can also give a direct proof of this latter direction of the corollary, as follows. We are given completely nonunitary row contractions $\mathbf{T} = (\mathbf{T}_1, \dots, \mathbf{T}_d)$ and $\mathbf{T}' = (\mathbf{T}'_1, \dots, \mathbf{T}'_d)$ such that $(T, L) := (T_{\mathbf{T}}, L_{\mathbf{T}})$ and $(T', L') := (T_{\mathbf{T}'}, L_{\mathbf{T}'})$ coincide. To simplify the notation, let us write the Halmos unitary dilations U and U' for $\mathbf{T}$ and $\mathbf{T}'$ in the form

$$U = \begin{bmatrix} A & B \\ C & D \end{bmatrix} : \begin{bmatrix} \mathcal{H} \\ \mathcal{E} \end{bmatrix} \to \begin{bmatrix} \oplus_{j=1}^d \mathcal{H} \\ \mathcal{E}_* \end{bmatrix}, \qquad U' = \begin{bmatrix} A' & B' \\ C' & D' \end{bmatrix} : \begin{bmatrix} \mathcal{H}' \\ \mathcal{E}' \end{bmatrix} \to \begin{bmatrix} \oplus_{j=1}^d \mathcal{H}' \\ \mathcal{E}'_* \end{bmatrix}$$

By assumption there are unitary transformations $\iota \colon \mathcal{E} \to \mathcal{E}'$ and $\iota_* \colon \mathcal{E}_* \to \mathcal{E}'_*$ so that

$$T'(z)\iota = \iota_* T(z), \qquad \widehat{L}'(z, \zeta)\iota = \iota \widehat{L}(z, \zeta). \tag{5.3.4}$$

On the other hand, from the formulas (5.3.1) and (5.3.2) we see that

$$T(z) = D + \sum_{j=1}^d \sum_{v \in \mathcal{F}_d} (C A^v B_j) z^v z_j,$$

$$\widehat{L}(z, \zeta) = \sum_{k,j=1}^d \sum_{v,w \in \mathcal{F}_d} B_k^* A^{*w} A^v B_j z^v z_j \zeta_k \zeta^w + I - T(\zeta)^* T(z) - T(\zeta)^* k_{per}(z, \zeta) T(z),$$

$$T'(z) = D' + \sum_{j=1}^d \sum_{v \in \mathcal{F}_d} (C' A'^v B'_j) z^v z_j,$$

$$\widehat{L}(z, \zeta) = \sum_{k,j=1}^d \sum_{v,w \in \mathcal{F}_d} B_k'^* A'^{*w} A'^v B'_j z^v z_j \zeta_k \zeta^w + I - T(\zeta)'^* T'(z) - T(\zeta)'^* k_{per}(z, \zeta) T'(z). \tag{5.3.5}$$

Combine (5.3.4) and (5.3.5) to see that
(5.3.6)
$$\iota_*^* D' \iota = D, \qquad \iota_*^* C' A'^v B_k' \iota = C A^v B_k, \qquad \iota^* B_k'^* A'^{*w} A'^v B_j' \iota = B_k{}^* A^{*w} A^v B_j$$

for all $v, w \in \mathcal{F}_d$ and $k, j = 1, \ldots, d$. The identities (5.3.6) suggest that we define an operator V with domain

$$\mathcal{D} = \{A^{*\alpha} A^\beta B_j e + A^{*v} C^* e_* \colon \alpha, \beta, v \in \mathcal{F}_d,\ j = 1, \ldots, d,\ e \in \mathcal{E},\ e_* \in \mathcal{E}_*\}$$

and range

$$\mathcal{D}' = \{A'^{*\alpha} A'^\beta B_j' e' + A'^{*v} C'^* e_{\prime *} \colon \alpha, \beta, v \in \mathcal{F}_d,\ j = 1, \ldots, d,\ e_* \in \mathcal{E}',\ e_*' \in \mathcal{E}_*'\}$$

by

$$V \colon A^{*\alpha} A^\beta B_j e + A^{*v} C^* e_* \mapsto A'^{*\alpha} A'^\beta B_j' \iota e + A'^{*v} C'^* \iota_* e_*. \tag{5.3.7}$$

We claim that V is isometric from $\mathcal{D}$ to $\mathcal{D}'$. For this to be true, we need

$$\begin{bmatrix} B_k^* A^{*\beta'^\top} A^{\alpha'^\top} \\ C A^{v'^\top} \end{bmatrix} \begin{bmatrix} A^{*\alpha} A^\beta B_j & A^{*v} C^* \end{bmatrix} = \begin{bmatrix} \iota^* B_k'^* A'^{*\beta'^\top} A'^{\alpha'^\top} \\ \iota_*^* C' A'^{v'^\top} \end{bmatrix} \begin{bmatrix} A'^{*\alpha} A'^\beta B_j' \iota & A'^{*v} C'^* \iota_* \end{bmatrix}$$

or, equivalently,

$$B_k^* A^{*\beta'^\top} A^{\alpha'^\top} A^{*\alpha} A^\beta B_j = \iota^* B_k'^* A'^{*\beta'^\top} A'^{\alpha'^\top} A'^{*\alpha} A^\beta B_j \iota, \tag{5.3.8}$$

$$B_k^* A^{*\beta'^\top} A^{\alpha'^\top} A^{*v} C^* = \iota^* B_k'^* A'^{*\beta'^\top} A'^{\alpha'^\top} A'^{*v} C'^* \iota_*, \tag{5.3.9}$$

$$C A^{v'^\top} A^{*v} C^* = \iota_*^* C' A'^{v'^\top} A'^{*v} C'^* \iota_*, \tag{5.3.10}$$

for all $\alpha, \beta, v, \alpha', \beta', v' \in \mathcal{F}_d$ and $k, j = 1, \ldots, d$. Note that particular cases of (5.3.8), (5.3.9) and (5.3.10) are already encoded in the known identities (5.3.6). To get the remaining cases of (5.3.8)–(5.3.10), use the identities

$$\begin{bmatrix} A_1^* & \cdots & A_d^* & C^* \\ B_1^* & \cdots & B_d^* & D^* \end{bmatrix} \begin{bmatrix} A_1 & B_1 \\ \vdots & \vdots \\ A_d & B_d \\ C & D \end{bmatrix} = \begin{bmatrix} I & 0 \\ 0 & I \end{bmatrix},$$

$$\begin{bmatrix} A_1 & B_1 \\ \vdots & \vdots \\ A_d & B_d \\ C & D \end{bmatrix} \begin{bmatrix} A_1^* & \cdots & A_d^* & C^* \\ B_1^* & \cdots & B_d^* & D^* \end{bmatrix} = \begin{bmatrix} I & & \\ & \ddots & \\ & & I \\ & & & I \end{bmatrix}$$

(and similarly for A', B', C' and D') together with an inductive argument on the length of the words $\alpha, \beta, v, \alpha', \beta', v'$ to deduce (5.3.8), (5.3.9) and (5.3.10) for the general case. We thus see that U is well-defined and isometric as an operator from $\mathcal{D}$ to $\mathcal{D}'$.

It then follows that V has a unique unitary extension from the closure of $\mathcal{D}$ in $\mathcal{H}$ to the closure of $\mathcal{R}$ in $\mathcal{H}'$. By Theorem 3.3.4 we know that $\mathcal{D}$ is dense in $\mathcal{H}$ exactly when $\mathbf{T}$ is completely nonunitary, and similarly, $\mathcal{D}'$ is dense in $\mathcal{H}'$ exactly when $\mathbf{T}'$ is completely nonunitary. Hence, under the assumption that both $\mathbf{T}$ and

$\mathbf{T}'$ are completely nonunitary, V extends to a unitary operator from $\mathcal{H}$ to $\mathcal{H}'$. From the definition of V and the identities (5.3.6) we get the intertwining property

$$\begin{bmatrix} V & 0 \\ 0 & \beta \end{bmatrix} \begin{bmatrix} A & B \\ C & D \end{bmatrix} = \begin{bmatrix} A' & B' \\ C' & D' \end{bmatrix} \begin{bmatrix} V & 0 \\ 0 & \alpha \end{bmatrix}.$$

In particular, V implements a unitary equivalence between A and A', and hence also between $\mathbf{T}$ and $\mathbf{T}'$. This concludes the direct proof of the nontrivial direction $\Longrightarrow$ and completes the proof of Corollary 5.3.3. □

REMARK 5.3.4. In case $d = 1$, the third set of conditions in (5.3.6) is already implied by the first two sets of conditions together with the unitary property of U. In this way we arrive at the proof found e.g. in [**BaC91**] that two closely-connected (1-variable) unitary colligations with the same characteristic function (or completely nonunitary contraction operators with the same characteristic function) are unitarily equivalent.

REMARK 5.3.5. For $\mathbf{T} = (\mathbf{T}_1, \dots, \mathbf{T}_d)$ a row contraction on a Hilbert space $\mathcal{H}$, let us say that $\mathbf{T}$ is *completely noncoisometric* (c.n.c.) if there is no nonzero invariant subspace for $\mathbf{T}_1^*, \dots, \mathbf{T}_d^*$ on which $\mathbf{T}_1\mathbf{T}_1^* + \cdots + \mathbf{T}_d\mathbf{T}_d^* = I$. In terms of the structure of the minimal row-isometric dilation $\mathcal{S} = (\mathcal{S}_1, \dots, \mathcal{S}_d)$ for $\mathbf{T}$ acting on the space $\mathcal{K}^+$ with Wold decomposition

$$\mathcal{K}^+ = \oplus_{v \in \mathcal{F}_d} \mathcal{S}^v \mathcal{E}_* \oplus \mathcal{R}$$

(where $\mathcal{E}_* := \mathcal{K}^+ \ominus \operatorname{span}_{j=1,\dots,d} \mathcal{S}_j \mathcal{K}^+$), this condition translates to $\mathbf{T}$ *is c.n.c.* $\Longleftrightarrow$ $\mathcal{R}_{00} := P_{\mathcal{R}}\mathcal{G}$ *is dense in* $\mathcal{R}$ where $\mathcal{G} = \mathcal{K}^+ \ominus \mathcal{H}$. In this case we may replace the dense set $\mathcal{D}_0$ given by (5.1.4) by the smaller dense set

$$\mathcal{D}_{00} := \mathcal{G}_*^+ + \mathcal{R}_{00} = \{k_1 + P_{\mathcal{R}}k_2 \colon k_1 \in \mathcal{G}_*^+,\ k_2 \in \mathcal{G}\}.$$

As $\Phi_{\mathcal{R}}$ (defined by (5.1.7)) is unitary from $\mathcal{R}$ onto $\mathcal{L}_L$ and by definition $\Phi_{\mathcal{R}}\mathcal{R}_{00} = L \cdot L^2(\mathcal{F}_d, \mathcal{E})$, the fact that $\mathcal{R}_{00}$ is dense in $\mathcal{R}$ implies that $L \cdot L^2(\mathcal{F}_d, \mathcal{E})$ is dense in $\mathcal{L}_L$, i.e., $\mathcal{H}_L = \mathcal{L}_L$. As $L^+ = I - (M_T)^* M_T$, from Theorem 2.4.6 we see that we are in the case where $I - (M_T)^* M_T$ has maximal factorable minorant equal to 0, or equivalently, the Cuntz-weight extension L of $I - (M_T)^* M_T$ is unique. Thus the characteristic pair (T, L) can be collapsed to simply the characteristic function T. As we saw from the calculation (2.3.1) in Section 2.3, the map

$$f(z, \zeta) = \sum_{v, w \in \mathcal{F}_d} f_{v,w} z^v \zeta^w \mapsto f(z, 0) := \sum_{v \in \mathcal{F}_d} f_{v,\emptyset} z^v$$

is a unitary transformation from $\mathcal{L}_L = \mathcal{H}_L$ onto $\mathcal{H}_{L^+}^+ = \mathcal{H}_{I-(M_T)^* M_T}^+$. Implementing this identification, we may take the Sz.-Nagy-Foiaş model to have the form

$$\mathcal{H}_T := \begin{bmatrix} L^2(\mathcal{F}_d, \mathcal{E}_*) \\ \mathcal{H}_{I-(M_T)^* M_T}^+ \end{bmatrix} \ominus \begin{bmatrix} T \\ I - (M_T)^* M_T \end{bmatrix} L^2(\mathcal{F}_d, \mathcal{E})$$

with associated model row-isometry $\mathbf{T}_T = (\mathbf{T}_{T,1}, \dots, \mathbf{T}_{T,d})$ given by

$$\mathbf{T}_{T,j} = P_{\mathcal{H}_T} \begin{bmatrix} S_j^R & 0 \\ 0 & \mathcal{S}_{(I-(M_T)^* M_T), j} \end{bmatrix} \Big|_{\mathcal{H}_T^+}.$$

In this way, for the c.n.c. case, we are able to do a complete Sz.-Nagy-Foiaş model theory without any reliance on the model theory of row-unitary d-tuples and Cuntz weights; this is essentially the model theory presented in detail in the work of Popescu [**Po89a, Po89b**].

REMARK 5.3.6. A partial unitary invariant for a row contraction

$$\mathbf{T} = (\mathbf{T}_1, \dots, \mathbf{T}_d)$$

with $\dim \mathcal{D}_{\mathbf{T}} < \infty$ called the *curvature* of $\mathbf{T}$ (analogous to Arveson's curvature invariant—see [**Arv00**]) for the case of a commuting d-tuple $\mathbf{T}$) has been introduced by Kribs [**Kr01**] and Popescu [**Po01a**]. This invariant $\mathrm{curv}(\mathbf{T})$ can be defined either as

$$\mathrm{curv}(\mathbf{T}) = (d-1) \lim_{k \to \infty} \frac{\mathrm{tr}(I - C_{\mathbf{T}}^k(I))}{d^k} \tag{5.3.11}$$

where $C_{\mathbf{T}}$ is the completely positive map on $\mathcal{L}(\mathcal{H})$ given by

$$C_{\mathbf{T}}(X) = \sum_{k=1}^{d} \mathbf{T}_j X \mathbf{T}_j^*,$$

or as

$$\mathrm{curv}(\mathbf{T}) = \lim_{m \to \infty} \frac{1}{d^m} \mathrm{tr}\left((P_m \otimes I)\Phi_*(\mathbf{T})\Phi_*(\mathbf{T})^*(P_m \otimes I)\right) \tag{5.3.12}$$

where $P_m \otimes I \colon L^2(\mathcal{F}_d, \mathcal{D}_{\mathbf{T}^*}) \to L^2(\mathcal{F}_d, \mathcal{D}_{\mathbf{T}^*})$ is the series-truncation projection operator

$$P_m \otimes I \colon \sum_{v \in \mathcal{F}_d} f_v z^v \mapsto \sum_{v \in \mathcal{F}_d \colon |v| = m} f_v z^v$$

and $\Phi_*(\mathbf{T}) \colon \mathcal{H} \to L^2(\mathcal{F}_d, \mathcal{D}_{\mathbf{T}^*})$ is the restriction to the scattering subspace $\mathcal{H}$ of the Fourier representation operator Φ_* associated with a Cuntz Lax-Phillips scattering system dilating $\mathbf{T}$: [1]

$$\Phi_*(\mathbf{T}) \colon h \mapsto \sum_{v \in \mathcal{F}_d} (D_{\mathbf{T}^*} \mathbf{T}^{*v} h) z^v$$

As $\mathrm{curv}(\mathbf{T})$ is a partial unitary invariant and the characteristic pair $(T_{\mathbf{T}}, L_{\mathbf{T}})$ is a complete unitary invariant for $\mathbf{T}$, it follows that one must be able to express $\mathrm{curv}(\mathbf{T})$ directly in terms of the characteristic pair $(T_{\mathbf{T}}, L_{\mathbf{T}})$. Explicitly, if we identify $\mathbf{T}$ with its Sz.-Nagy-Foiaş model given as in Theorem 5.3.1 (with $(T, L) = (T_{\mathbf{T}}, L_{\mathbf{T}})$), then the minimal row-isometric dilation $\mathcal{U}_+ = (\mathcal{U}_{+1}, \dots, \mathcal{U}_{+d})$ of $\mathbf{T}$ has the form

$$\mathcal{U}_{+j} = \begin{bmatrix} S_j^R & 0 \\ 0 & \mathcal{U}_{L,j} \end{bmatrix} \text{ on } \begin{bmatrix} L^2(\mathcal{F}_d, \mathcal{D}_{\mathbf{T}^*}) \\ \mathcal{L}_L \end{bmatrix}$$

where we take $\mathbf{T} = (\mathbf{T}_1, \dots, \mathbf{T}_d)$ to be equal to its model

$$\mathbf{T}_j = P_{\mathcal{H}} \begin{bmatrix} S_j^R & 0 \\ 0 & \mathcal{U}_{L,j} \end{bmatrix}\Bigg|_{\mathcal{H}} \text{ on } \mathcal{H} = \begin{bmatrix} L^2(\mathcal{F}_d, \mathcal{D}_{\mathbf{T}^*}) \\ \mathcal{L}_L \end{bmatrix} \ominus \begin{bmatrix} T \\ L \end{bmatrix} L^2(\mathcal{F}_d, \mathcal{D}_{\mathbf{T}}).$$

Then (5.3.12) can be rewritten in the form

$$\mathrm{curv}(\mathbf{T}) = \frac{1}{d^m} \sum_{v \colon |v| = m} \mathrm{tr}\left(P_{\mathcal{H}} \mathcal{U}_+^v (I - \sum_{j=1}^{d} \mathcal{U}_{+j} \mathcal{U}_{+j}^*) \mathcal{U}_+^{*v^\top} P_{\mathcal{H}} \right). \tag{5.3.13}$$

where the orthogonal projection $P_{\mathcal{H}}$ onto $\mathcal{H}$ has the explicit form

$$P_{\mathcal{H}} = \begin{bmatrix} I_{L^2(\mathcal{F}_d, \mathcal{D}_{\mathbf{T}^*})} & 0 \\ 0 & I_{\mathcal{L}_L} \end{bmatrix} - \begin{bmatrix} M_T \\ L \end{bmatrix} \begin{bmatrix} M_{\mathbf{T}}^* & L \end{bmatrix} \text{ on } \begin{bmatrix} L^2(\mathcal{F}_d, \mathcal{D}_{\mathbf{T}^*}) \\ \mathcal{L}_L \end{bmatrix} \tag{5.3.14}$$

[1] If one replaces tr with rank in either (5.3.11) or (5.3.12), one gets the related invariant called the *Euler characteristic* which plays a role in an operator-theoretic analogue of the Gauss-Bonnet Theorem—see [**Arv00**]; for the sake of conciseness, we focus here on the curvature invariant.

Substitution of (5.3.14) into (5.3.13) then gives

$$\text{curv}(\mathbf{T}) = \dim \mathcal{D}_{\mathbf{T}^*} - \lim_{m\to\infty} \frac{1}{d^m} \sum_{v\in\mathcal{F}_d\colon |v|\le m} \text{tr}(T_v^* T_v) \tag{5.3.15}$$

where T_v ($v \in \mathcal{F}_d$) are the Taylor coefficients of the characteristic function $T_{\mathbf{T}}(z) = \sum_{v\in\mathcal{F}_d} T_v z^v$, i.e.,

$$T_\emptyset = -\mathbf{T}|_{\mathcal{D}_{\mathbf{T}}}, \qquad T_{g_{i_n}\cdots g_{i_1}} = D_{\mathbf{T}^*}\mathbf{T}^*_{i_N}\cdots \mathbf{T}^*_{i_2}[D_{\mathbf{T}}]_{i_1}|_{\mathcal{D}_{\mathbf{T}}}$$

(where $[D_{\mathbf{T}}]_j\colon \oplus_{k=1}^m \mathcal{H} \to \mathcal{H}$ refers to the j-th row of $D_{\mathbf{T}}$ in its matrix decomposition as an operator on $\oplus_{j=1}^d \mathcal{H}$). In particular we see that the curvature invariant curv$(\mathbf{T})$ is completely determined by the characteristic function $T_{\mathbf{T}}$ alone and does not see the characteristic Cuntz weight $L_{\mathbf{T}}$.

The idea of the curvature invariant curv$(\mathbf{T})$ is that, while it is only a partial unitary invariant, it is in general easier to compute than the complete invariant $(T_{\mathbf{T}}, L_{\mathbf{T}})$ and is still fine enough to distinguish some properties of $\mathbf{T}$. For example, it is shown in [**Kr01, Po01a**] that, *given that* $\mathbf{T}$ *is of class* $C_{\cdot 0}$, i.e.,

$$\lim_{N\to\infty} \sum_{v\in\mathcal{F}_d\colon |v|=N} T^v T^{*v^\top} h = 0 \text{ for each } h \in \mathcal{H},$$

then $\mathbf{T}$ *is unitarily equivalent to the row shift* $S^R = (S_1^R, \dots, S_d^R)$ *on* $L^2(\mathcal{F}_d, \mathcal{E}_*)$ *for some coefficient space* $\mathcal{E}_*$ *if and only if* curv$(\mathbf{T})$ = rank $D_{\mathbf{T}^*}$ (and then $\dim \mathcal{E}_* = \text{rank}\, D_{\mathbf{T}^*}$). Indeed, the characteristic pair (T, L) for the row shift is given by $T\colon \{0\} \to \mathcal{E}_*$ and $L = 0\colon \{0\} \to \{0\}$. From the formula (5.3.15) for curv$(\mathbf{T})$, we see that curv$(\mathbf{T})$ = rank $D_{\mathbf{T}^*}$ (= $\dim \mathcal{D}_{\mathbf{T}^*}$) if and only if $T(z) = 0\colon L^2(\mathcal{F}_d, \mathcal{E}) \to L^2(\mathcal{F}_d, \mathcal{E}_*)$. Under the assumption that $\mathbf{T}$ is $C_{\cdot 0}$, it is not hard to see that in fact $\mathcal{E} = \{0\}$, and hence $(T, L) = (T_{S^R}, L_{S^R})$, forcing $\mathbf{T}$ to be unitarily equivalent to the row shift S^R on $L^2(\mathcal{F}_d, \mathcal{E}_*)$ as expected.

5.4. Realization theory

We have seen that any Cuntz unitary colligation

$$U = \begin{bmatrix} A & B \\ C & D \end{bmatrix} = \begin{bmatrix} A_1 & B_1 \\ \vdots & \vdots \\ A_d & B_d \\ C & D \end{bmatrix} \colon \begin{bmatrix} \mathcal{H} \\ \mathcal{E} \end{bmatrix} \to \begin{bmatrix} \oplus_{j=1}^d \mathcal{H} \\ \mathcal{E}_* \end{bmatrix}, \qquad U \text{ unitary} \tag{5.4.1}$$

has a characteristic function given by

$$T(z) = D + C(I - Z_r(z)A)^{-1} Z_r(z) B \in \mathcal{S}_{nc,d}(\mathcal{E}, \mathcal{E}_*). \tag{5.4.2}$$

In this section, we treat the converse question: *given a* $T(z) \in \mathcal{S}_{nc,d}(\mathcal{E}, \mathcal{E}_*)$, *is there a Cuntz unitary colligation* $U = \begin{bmatrix} A & B \\ C & D \end{bmatrix}$ *so that* $T(z)$ *can be "realized" in the form* (5.4.2)? More generally, one can pose the question without the requirement that U be unitary. For the $d = 1$ case, this is known as the "realization problem", and its solution has been a major tool in a number of different areas, e.g., in interpolation and factorization theory for rational and meromorphic operator-valued functions, and in the analysis of Wiener-Hopf and singular integral equations (see [**BarGK79, BaGR90, Ka85, Ka96**] as well as [**Ba00, BaY98**] for surveys and open questions concerning recent multivariable generalizations). It was observed some time ago by Helton [**H72**] that the Sz.-Nagy-Foiaş model theory is directly applicable to the problem of realizing a given Schur-class function as the transfer function of a unitary system.

It is the purpose of this section to extend the idea in [**H72**] to the present noncommutative, multivariable setting. An extra complication is the fact that the complete unitary invariant for a closely-connected unitary colligation (or c.n.u. row contraction) is the characteristic pair (T, L) rather than the characteristic function T alone. In any case, we have the following realization result.

THEOREM 5.4.1. *Let $T(z) \in \mathcal{S}_{nc,d}(\mathcal{E}, \mathcal{E}_*)$ be given. Then there is a closely-connected Cuntz unitary colligation $U = \begin{bmatrix} A & B \\ C & D \end{bmatrix}$ so that $T(z)$ is realized in the form* (5.4.2).

To construct such a U, choose any Cuntz-weight extension L of $I - (M_T)^* M_T$ (Theorem 2.4.7 gives a description of all such L in terms of the maximal factorable minorant $(L_{T_1})^* L_{T_1} \leq I - (M_T)^* M_T$ for $I - (M_T)^* M_T$). Then define U of the form (5.4.1) *with the state space $\mathcal{H}$ taken to be*

$$\mathcal{H}_{T,L} := \mathcal{H}_{NF'} = \begin{bmatrix} L^2(\mathcal{F}_d, \mathcal{E}_*) \\ \mathcal{L}_L \end{bmatrix} \ominus \begin{bmatrix} T \\ L \end{bmatrix} L^2(\mathcal{F}_d, \mathcal{E})$$

and with operators $A_j \colon \mathcal{H}_{T,L} \to \mathcal{H}_{T,L}$, $B_j \colon \mathcal{E} \to \mathcal{H}_{T,L}$, $C \colon \mathcal{H}_{T,L} \to \mathcal{E}_*$ and $D \colon \mathcal{E} \to \mathcal{E}_*$ given by

$$A_j = P_{\mathcal{H}_{T,L}} \begin{bmatrix} S_j^{R[*]} & 0 \\ 0 & (\mathcal{U}_{L,j})^{[*]} \end{bmatrix} \Bigg|_{\mathcal{H}_{T,L}} \quad \text{for } j = 1, \dots, d,$$

$$B_j \colon e \mapsto P_{\mathcal{H}_{T,L}} \begin{bmatrix} S_j^{R[*]} & 0 \\ 0 & (\mathcal{U}_{L,j})^* \end{bmatrix} e \text{ for } e \in \mathcal{E} \text{ and } j = 1, \dots, d,$$

$$C \colon \begin{bmatrix} f \\ g \end{bmatrix} \mapsto f_\emptyset \text{ for } \begin{bmatrix} f \\ g \end{bmatrix} \in \mathcal{H}_{T,L} \text{ with } f(z) = \sum_{v \in \mathcal{F}_d} f_v z^v,$$

$$D \colon e \mapsto T_\emptyset e.$$

PROOF. Given $T(z) \in \mathcal{S}_{nc,d}(\mathcal{E}, \mathcal{E}_*)$, we can always find a Cuntz weight extension L of $I - (M_T)^* M_T$ by the procedure in Theorem 2.4.7, so that (T, L) is a characteristic pair. The formula for U in the statement of the theorem amounts to the model Cuntz unitary colligation U' associated with the Sz.-Nagy-Foiaş-model outgoing Cuntz scattering system $\mathfrak{S}_{NF'}^+$ (5.1.10) constructed from (T, L), with the adjustment that the identification operators $i_{NF'}$ and $i_{NF'*}$ are used to identify $\mathcal{E}$ with $\mathcal{E}_{NF'}$ and $\mathcal{E}_*$ with $\mathcal{E}_{NF'*}$ respectively. By Theorem 5.1.1, the characteristic function of $\mathfrak{S}_{NF'}^+$, which is the same as the characteristic function of the colligation U', coincides (via $i_{NF'}$ and $i_{NF'*}$) with $T(z)$, i.e., the characteristic function of U is equal to $T(z)$. The Theorem now follows. □

Due to the flexibility in the choice of L in Theorem 5.4.1, we cannot expect the closely-connected Cuntz unitary colligation which realizes a given $T(z)$ to be uniquely determined up to unitary equivalence except in the case where $I - (M_T)^* M_T$ has zero maximal factorable minorant. If we also specify the auxiliary symbol $X(z, \zeta)$ given by (4.2.51), then we get a realization result which includes a uniqueness statement.

THEOREM 5.4.2. *Suppose that $T(z)$ and $X(z,\zeta)$ are formal power series of the respective forms*

$$T(z) = \sum_{v \in \mathcal{F}_d} T_v z^v, \qquad X(z,\zeta) = \sum_{i,j=1}^{d} \zeta_j X^{ij}(z,\zeta) z_i.$$

Then there is a closely-connected Cuntz unitary colligation U of the form (5.4.1) such that

$$T(z) = D + C(I - Z_r(z)A)^{-1} Z_r(z) B \tag{5.4.3}$$

$$X(z,\zeta) = B^* Z_r(\zeta)^* (I - A^* Z_r(\zeta)^*)^{-1} (I - Z_r(z)A)^{-1} Z_r(z) B \tag{5.4.4}$$

if and only if

(1) *$X(z,\zeta)$ is a positive symbol, and*
(2) *$X(z,\zeta)$ and $T(z)$ satisfy the identity*

$$I - T(\zeta)^* T(z) = \sum_{j=1}^{d} z_j^{-1} X(z,\zeta) \zeta_j^{-1} - X(z,\zeta). \tag{5.4.5}$$

In this case, if $U' = \begin{bmatrix} \mathcal{H}' \\ \mathcal{E} \end{bmatrix} \to \begin{bmatrix} \oplus_{j=1}^{d} \mathcal{H}' \\ \mathcal{E}_* \end{bmatrix}$ is another such closely-connected Cuntz unitary colligation, then there is a unitary transformation $i\colon \mathcal{H} \to \mathcal{H}'$ so that

$$\begin{bmatrix} (\oplus_{j=1}^{d} i) & 0 \\ 0 & I \end{bmatrix} \begin{bmatrix} A & B \\ C & D \end{bmatrix} = \begin{bmatrix} A' & B' \\ C' & D' \end{bmatrix} \begin{bmatrix} i & 0 \\ 0 & I \end{bmatrix}.$$

PROOF. If U is a Cuntz unitary colligation for which (5.4.3) and (5.4.4) hold, by Proposition 4.2.11, it follows that

$$\widehat{L}(z,\zeta) = X(z,\zeta) + I - T(\zeta)^* T(z) - T(\zeta)^* k_{per}(z,\zeta) T(z) \tag{5.4.6}$$

is the symbol for a Cuntz-weight extension L of $I - (M_T)^* M_T$. Then, by Theorem 2.4.5, it follows that, in addition to $X(z,\zeta)$ being a positive symbol, we must have that $X'(z,\zeta) = 0$, where

$$X'(z,\zeta) := X(z,\zeta) - \sum_{j=1}^{d} z_j^{-1} X(z,\zeta) \zeta_j^{-1} + I - T(\zeta)^* T(z) = 0$$

is as in (2.4.24). Thus, conditions (1) and (2) in the statement of the Theorem hold.

Conversely, suppose that $T(z)$ and $X(z,\zeta)$ are formal power series satisfying conditions (1) and (2) in the statement of the Theorem. Let W_* be any Cuntz-weight extension of $[\delta_{v,\alpha} I_{\mathcal{E}_*}]_{v,\alpha \in \mathcal{F}_d}$, and let W be the Haplitz operator with symbol given by (2.4.25):

$$\widehat{W}(z,\zeta) = X(z,\zeta) + I + T(\zeta)^* [\widehat{W}_*(z,\zeta) - I] T(z).$$

From our assumptions, we see that the conditions of part (2) of Theorem 2.4.5 hold with $X'(z,\zeta) = 0$. Hence W is a Cuntz-weight extension of $[\delta_{w,\alpha} I_{\mathcal{E}}]_{v,\alpha \in \mathcal{F}_d}$ and $L_T^{W,W_*}\colon \mathcal{L}_W \to \mathcal{L}_{W_*}$ is contractive. In particular, $M_T\colon L^2(\mathcal{F}_d,\mathcal{E}) \to L^2(\mathcal{F}_d,\mathcal{E}_*)$ is contractive and $T \in \mathcal{S}_{nc,d}(\mathcal{E},\mathcal{E}_*)$. A simple consequence of the definitions is that we recover X from (T, W, W_*) as in formula (2.4.16). Moreover, by part (3) of Theorem 2.4.5, we see that $\widehat{L}(z,\zeta)$ defined by (5.4.6) is the symbol for a Cuntz-weight extension of $I - (M_T)^* M_T$. We then construct the model Cuntz unitary colligation U for the characteristic pair (T, L) as in Theorem 5.4.1. One can now use Theorem 5.1.1 and Proposition 4.2.11 to see that T and X have the realized form (5.4.3) and (5.4.4) as desired.

Uniqueness follows from the uniqueness statement in Theorem 5.1.1. Alternatively, one could do a direct proof as in the proof of Corollary 5.3.3. This completes the proof of Theorem 5.4.2. □

Bibliography

[AA95] V.M. Adamjan and D.Z. Arov, *On unitary coupling of semiunitary operators*, Dokl. Akad. Nauk. Arm. SSR XLIII, **5** (1966), 257-263 [in Russian] = Amer. Math. Soc. Transl. **95** (1970), 75-129.

[AFMP94] G.T. Adams, J. Froelich, P.J. McGuire and V.I. Paulsen, *Analytic reproducing kernels and factorization*, Indiana Univ. Math. J. **43** (1994), 839-856.

[AM00] J. Agler and J.E. McCarthy, *Complete Nevanlinna-Pick kernels*, J. Functional Analysis **175** (2000), 111-124.

[ADRS97] D. Alpay, A. Dijksma, J.R. Rovnyak and H. de Snoo, *Schur Functions, Operator Colligations, and Reproducing Kernel Pontryagin Spaces*, OT 96, Birkhäuser-Verlag, Basel-Boston, 1997.

[AP00] A. Arias and G. Popescu, *Noncommutative interpolation and Poisson transforms*, Israel J. Math. **115** (2000), 205-234.

[Arv98] W. Arveson, *Subalgebras of C^*-algebras III: multivariable operator theory*, Acta Math. **181** (1998), 159-228.

[Arv00] W. Arveson, *The curvature invariant of a Hilbert module over $C[z_1, \dots, z_d]$*, J. Reine Angew. Math. **522** (2000), 173-236.

[Ba00] J.A. Ball, *Linear systems, operator model theory and scattering: multivariable generalizations*, Operator Theory and Its Applications (Winnipeg, MB, 1998) (Ed. A.G. Ramm, P.N. Shivakumar and A.V. Strauss), Fields Institute Communications Vol. 25, Amer. Math. Soc., Providence, 2000, pp. 151-178

[BaC91] J.A. Ball and N. Cohen, *De Branges-Rovnyak operator models and systems theory: a survey*, Topics in Matrix and Operator Theory (Ed. H. Bart, I. Gohberg and M.A. Kaashoek), OT 50 Birkhäuser-Verlag, Basel-Boston, 1991, pp. 93-136.

[BaGR90] J.A. Ball, I. Gohberg and L. Rodman, *Interpolation of Rational Matrix Functions*, OT45 Birkhäuser-Verlag, Basel-Boston, 1990.

[BaT98] J.A. Ball and T.T. Trent, *Unitary colligations, reproducing kernel Hilbert spaces, and Nevanlinna-Pick interpolation in several variables*, J. Functional Analysis **157** (1998), 1-61.

[BaT00] J.A. Ball and T.T. Trent, *The abstract interpolation problem and commutant lifting: A coordinate-free approach*, Operator Theory and Interpolation: International Workshop on Operator Theory and Applications, IWOTA96 (Ed. H. Bercovici and C. Foiaş), OT115 Birkhäuser-Verlag, Basel-Boston, 2000, pp. 51-83

[BaTV01] J.A. Ball, T.T. Trent and V. Vinnikov, *Interpolation and commutant lifting for multipliers on reproducing kernel Hilbert spaces with embedding into a ball*, Operator Theory and Analysis: The M.A. Kaashoek Anniversary Volume: Workshop in Amsterdam, Nov. 1997 (Ed. H. Bart, I. Gohberg and A.C.M. Ran), OT 122 Birkhäuser-Verlag, Basel-Boston, 2001, pp. 89-138.

[BaV03] J.A. Ball and V. Vinnikov, *Formal reproducing kernel Hilbert spaces: the commutative and noncommutative settings*, Reproducing Kernel Spaces and Applications (Ed. D. Alpay), OT143 Birkhäuser-Verlag, Basel-Boston, 2003, pp. 77-134

[BaV04] J.A. Ball and V. Vinnikov, *Functional models for representations of the Cuntz algebra*, Operator Theory, System Theory and Scattering Theory: Multidimensional Generalizations (Ed. D. Alpay and V. Vinnikov), Birkhäuser-Verlag, Basel-Boston, in press.

[BaY98] J.A. Ball and N.J. Young, *Problems on the realization of functions*, Operator Theory and Its Applications (Winnipeg, MB, 1998) (Ed. A.G. Ramm, P.N. Shivakumar and A.V. Strauss), Fields Institute Communications Vol. 25, Amer. Math. Soc., Providence, 2000, pp. 151-178

[BarGK79] H. Bart, I. Gohberg and M.A. Kaashoek, *Minimal Factorization of Matrix and Operator Functions*, OT1 Birkhäuser-Verlag, Basel-Boston, 1979.

[Be01] C.L. Beck, *On formal power series representations for uncertain systems*, IEEE Transactions on Automatic Control **46** (2001), 314-319.

[BeD99] C.L. Beck and J. Doyle, *A necessary and sufficient minimality condition for uncertain systems*, IEEE Transactions on Automatic Control **44** (1999), 1802-1813.

[BoDK00] S.S. Boiko, V.K. Dubovoy and A.Ya. Kheifets, *Measure Schur complements and spectral functions of unitary operators with respect to different scales*, Operator Theory, System Theory and Related Topics: The Moshe Livšic Anniversary Volume (Ed. D. Alpay and V. Vinnikov), OT 123 Birkhäuser, Basel-Boston, 2000, pp. 89-138.

[BraR66] L. de Branges and J. Rovnyak, *Canonical models in quantum scattering theory*, Perturbation Theory and its Applications in Quantum Mechanics (Ed. C.H. Wilcox), Wiley, New York, 1966, pp. 295-392.

[Bro71] M.S. Brodskiĭ, *Triangular and Jordan Representations of Linear Operators*, Translations of Mathematical Monographs Volume 32, American Mathematical Society, Providence, 1971.

[Bu84] J.W. Bunce, *Models for n-tuples of noncommuting operators*, J. Functional Analysis **57** (1984), 21-30.

[CSK02] T. Constantinescu, A.H. Sayed and T. Kailath, *Inverse scattering experiments, structured matrix inequalities, and tensor algebras*, Linear Alg. Appl. **343/344** (2002), 147-169.

[Da96] K.R. Davidson, *C^*-Algebras by Example*, Fields Institute Monograph 6, American Mathematical Society, Providence, 1996.

[DaP98a] K.R. Davidson and D.R. Pitts, *The algebraic structure of non-commutative analytic Toeplitz algebras*, Math. Ann. **311** (1998), 275-303.

[DaP98b] K.R. Davidson and D.R. Pitts, *Nevanlinna-Pick interpolation for non-commutative analytic Toeplitz algebra*, Integral Equations and Operator Theory **31** (1998), 321-337.

[DaP99] K.R. Davidson and D.R. Pitts, *Invariant subspaces and hyper-reflexivity for free semigroup algebras*, Proc. London Math. Soc. **78** (1999), 401-430.

[Dr78] S.W. Drury, *A generalization of von Neumann's inequality to the complex ball*, Proc. Amer. Math. Soc. **68** (1978), 300-304.

[Fl74] M. Fliess, *Matrices de Hankel*, J. Math. pures et appl. **53** (1974), 197-224.

[Fr82] A.E. Frazho, *Models for noncommuting operators*, J. Functional Analysis **48** (1982), 1-11.

[Fr84] A.E. Frazho, *Complements to models for noncommuting operators*, J. Functional Analysis **59** (1984), 445-461.

[GRS02] D. Greene, S. Richter and C. Sundberg, *The structure of inner multipliers on spaces with complete Nevanlinna Pick kernels*, J. Functional Analysis **194** (2002), 311-321.

[H72] J.W. Helton, *The characteristic functions of operator theory and electrical network realization*, Indiana Univ. Math. J. **22** (1972/73), 403-414.

[Ka85] M.A. Kaashoek, *Minimal factorization, linear systems and integral operators*, Operators and function theory (Lancaster, 1984), NATO Adv. Sci. Inst. Ser. C Math. Phy. Sci. 153, Reidel, Dordrecht, 1985, pp. 41-86

[Ka96] M.A. Kaashoek, *State space theory of rational matrix functions and application*, Lectures on Operator Theory and its Applications (Waterloo, ON, 1994), Fields Inst. Monogr., Amer. Math. Soc., Prov., RI, 1996, pp. 233-333

[Kr01] D.W. Kribs, *The curvature invariant of a non-commuting N-tuple*, Integral Equations and Operator Theory **41** (2001), 426-454.

[LaP67] P.D. Lax and R.S. Phillips, *Scattering Theory*, Academic Press, New York-London, 1967.

[Li73] M.S. Livšic, *Operators, Oscillations, Waves (Open Systems)*, Translations of Mathematical Monographs Volume Thirty-Four, American Mathematical Society, Providence, 1973.

[McT00] S. McCullough and T.T. Trent, *Invariant subspaces and Nevanlinna-Pick kernels*, J. Functional Analysis **178** (2000), 226-249.

[NaF70] B. Sz.-Nagy and C. Foiaş, *Harmonic Analysis of Operators on Hilbert Space*, North Holland/American Elsevier, Amsterdam-New York, 1970.

[NiV89] N.K. Nikolskii and V.I Vasyunin, *A unified approach to function models, and the transcription problem*, The Gohberg Anniversary Collection Volume II: Topics in Analysis and Operator Theory (Ed. H. Dym, S. Goldberg, M.A. Kaashoek and P. Lancaster), OT41 Birkhäuser-Verlag, Basel-Boston, 1989, pp. 405-434.

[NiV98] N.K. Nikolskii and V.I. Vasyunin, *Elements of spectral theory in terms of the free function model Part I: Basic constructions*, Holomorphic Spaces (Ed. S. Axler, J.E. McCarthy and D. Sarason), Mathematical Sciences Research Institute Publications Vol. 33, Cambridge University Press, 1998, pp. 211-302.

[Po89a] G. Popescu, *Models for infinite sequences of noncommuting operators*, Acta Sci. Math. **53** (1989), 355-368.

[Po89b] G. Popescu, *Characteristic functions for infinite sequences of noncommuting operators*, J. Operator Theory **22** (1989), 51-71.

[Po89c] G. Popescu, *Isometric dilations for infinite sequences of noncommuting operators*, Trans. Amer. Math. Soc. **316** (1989), 523-536.

[Po91] G. Popescu, *Von Neumann inequality for* $(B(\mathcal{H}^n)_1$, Math. Scand. **68** (1991), 292-304.

[Po95] G. Popescu, *Multi-analytic operators on Fock spaces*, Math. Ann. **303** (1995), 31-46.

[Po98] G. Popescu, *Interpolation problems in several variables*, J. Math. Anal. Appl. **227** (1998), 227-250.

[Po99] G. Popescu, *Poisson transforms on some* C^**-algebras generated by isometries*, J. Functional Analysis **161** (1999), 27-61.

[Po01a] G. Popescu, *Curvature invariant for Hilbert modules over free semigroup algebras*, Adv. Math. **158** (2001), 264-309.

[Po01b] G. Popescu, *Structure and entropy for positive-definite Toeplitz kernels on free semigroups*, J. Math. Anal. Appl. **254** (2001), 191-218.

[Sh55] J.J. Schäffer, *On unitary dilations of contractions*, Proc. Amer. Math. Soc. **6** (1955), 322.

Editorial Information

To be published in the *Memoirs*, a paper must be correct, new, nontrivial, and significant. Further, it must be well written and of interest to a substantial number of mathematicians. Piecemeal results, such as an inconclusive step toward an unproved major theorem or a minor variation on a known result, are in general not acceptable for publication. Papers appearing in *Memoirs* are generally at least 80 and not more than 200 published pages in length. Papers less than 80 or more than 200 published pages require the approval of the Managing Editor of the Transactions/Memoirs Editorial Board.

As of July 31, 2005, the backlog for this journal was approximately 14 volumes. This estimate is the result of dividing the number of manuscripts for this journal in the Providence office that have not yet gone to the printer on the above date by the average number of monographs per volume over the previous twelve months, reduced by the number of volumes published in four months (the time necessary for preparing a volume for the printer). (There are 6 volumes per year, each containing at least 4 numbers.)

A Consent to Publish and Copyright Agreement is required before a paper will be published in the *Memoirs*. After a paper is accepted for publication, the Providence office will send a Consent to Publish and Copyright Agreement to all authors of the paper. By submitting a paper to the *Memoirs*, authors certify that the results have not been submitted to nor are they under consideration for publication by another journal, conference proceedings, or similar publication.

Information for Authors

Memoirs are printed from camera copy fully prepared by the author. This means that the finished book will look exactly like the copy submitted.

The paper must contain a *descriptive title* and an *abstract* that summarizes the article in language suitable for workers in the general field (algebra, analysis, etc.). The *descriptive title* should be short, but informative; useless or vague phrases such as "some remarks about" or "concerning" should be avoided. The *abstract* should be at least one complete sentence, and at most 300 words. Included with the footnotes to the paper should be the 2000 *Mathematics Subject Classification* representing the primary and secondary subjects of the article. The classifications are accessible from `www.ams.org/msc/`. The list of classifications is also available in print starting with the 1999 annual index of *Mathematical Reviews*. The Mathematics Subject Classification footnote may be followed by a list of *key words and phrases* describing the subject matter of the article and taken from it. Journal abbreviations used in bibliographies are listed in the latest *Mathematical Reviews* annual index. The series abbreviations are also accessible from `www.ams.org/publications/`. To help in preparing and verifying references, the AMS offers MR Lookup, a Reference Tool for Linking, at `www.ams.org/mrlookup/`. When the manuscript is submitted, authors should supply the editor with electronic addresses if available. These will be printed after the postal address at the end of the article.

Electronically prepared manuscripts. The AMS encourages electronically prepared manuscripts, with a strong preference for AMS-LaTeX. To this end, the Society has prepared AMS-LaTeX author packages for each AMS publication. Author packages include instructions for preparing electronic manuscripts, the *AMS Author Handbook*, samples, and a style file that generates the particular design specifications of that publication series. Though AMS-LaTeX is the highly preferred format of TeX, author packages are also available in AMS-TeX.

Authors may retrieve an author package from e-MATH starting from `www.ams.org/tex/` or via FTP to `ftp.ams.org` (login as `anonymous`, enter username as password, and type `cd pub/author-info`). The *AMS Author Handbook* and the *Instruction Manual* are available in PDF format following the author packages link from `www.ams.org/tex/`. The author package can be obtained free of charge by sending email to pub@ams.org (Internet) or from the Publication Division, American Mathematical Society, 201 Charles St., Providence, RI 02904, USA. When requesting an author package, please specify $\mathcal{A}_{\mathcal{M}}\mathcal{S}$-LaTeX or $\mathcal{A}_{\mathcal{M}}\mathcal{S}$-TeX, Macintosh or IBM (3.5) format, and the publication in which your paper will appear. Please be sure to include your complete mailing address.

Sending electronic files. After acceptance, the source file(s) should be sent to the Providence office (this includes any TeX source file, any graphics files, and the DVI or PostScript file).

Before sending the source file, be sure you have proofread your paper carefully. The files you send must be the EXACT files used to generate the proof copy that was accepted for publication. For all publications, authors are required to send a printed copy of their paper, which exactly matches the copy approved for publication, along with any graphics that will appear in the paper.

TeX files may be submitted by email, FTP, or on diskette. The DVI file(s) and PostScript files should be submitted only by FTP or on diskette unless they are encoded properly to submit through email. (DVI files are binary and PostScript files tend to be very large.)

Electronically prepared manuscripts can be sent via email to pub-submit@ams.org (Internet). The subject line of the message should include the publication code to identify it as a Memoir. TeX source files, DVI files, and PostScript files can be transferred over the Internet by FTP to the Internet node e-math.ams.org (130.44.1.100).

Electronic graphics. Comprehensive instructions on preparing graphics are available at www.ams.org/jourhtml/graphics.html. A few of the major requirements are given here.

Submit files for graphics as EPS (Encapsulated PostScript) files. This includes graphics originated via a graphics application as well as scanned photographs or other computer-generated images. If this is not possible, TIFF files are acceptable as long as they can be opened in Adobe Photoshop or Illustrator. No matter what method was used to produce the graphic, it is necessary to provide a paper copy to the AMS.

Authors using graphics packages for the creation of electronic art should also avoid the use of any lines thinner than 0.5 points in width. Many graphics packages allow the user to specify a "hairline" for a very thin line. Hairlines often look acceptable when proofed on a typical laser printer. However, when produced on a high-resolution laser imagesetter, hairlines become nearly invisible and will be lost entirely in the final printing process.

Screens should be set to values between 15% and 85%. Screens which fall outside of this range are too light or too dark to print correctly. Variations of screens within a graphic should be no less than 10%.

Inquiries. Any inquiries concerning a paper that has been accepted for publication should be sent directly to the Electronic Prepress Department, American Mathematical Society, 201 Charles St., Providence, RI 02904, USA.

Editors

Titles in This Series